Oil in Troubled Waters

Perceptions, Politics, and the Battle Over
Offshore Drilling

SUNY Series in Environmental Public Policy

David W. Orr and Harlan Wilson, Editors

OIL IN TROUBLED WATERS

Perceptions, Politics, and the Battle Over
Offshore Drilling

WILLIAM R. FREUDENBURG
and
ROBERT GRAMLING

State University of New York Press

Published by
State University of New York Press, Albany

© 1994 State University of New York

For information, address the State University of New York Press,
State University Plaza, Albany, NY 12246

Library of Congress Cataloging-in-Publication Data

Freudenburg, William R.
    Oil in troubled waters: perceptions, politics, and the battle over
offshore drilling / by William R. Freudenburg and Robert Gramling.
        p.   cm. — (SUNY series in environmental public policy)
    Includes bibliographical references (p.).
    ISBN 0-7914-1881-2 (cloth : acid-free paper). — ISBN
0-7914-1882-0 (pbk. : acid-free paper)
        1. Offshore oil industry—Government policy—Louisiana—Citizen
participation.   2. Offshore oil industry—Government policy-
-California—Citizen participation.   3. Offshore oil industry-
-Environmental aspects—Louisiana.   4. Offshore oil industry-
-Environmental aspects—California.   I. Gramling, Robert, 1943– .
II. Title.   III. Series.
HD9565.F73   1994
333.8'232'09763—dc20   93-24947
                                                                        CIP

10   9   8   7   6   5   4   3   2   1

To Betty, Eldon, Anne, and Bob

CONTENTS

LIST OF ACRONYMS

BLM	U.S. Bureau of Land Management
CEQ	U.S. Council on Environmental Quality
C.F.R.	"Code of Federal Regulations"—the topical codification of the *regulations*, by federal agencies, as opposed to the *legislation* passed by Congress.
DEIS	Draft Environmental Impact Statement
EIS	(final) Environmental Impact Statement
EPA	U.S. Environmental Protection Agency
MMS	U.S. Minerals Management Service, Department of Interior
NEPA	National Environmental Policy Act of 1969, P.L. 94–190, 42 U.S.C. 4321 et seq.
OCS	"Outer Continental Shelf"—the sea-bottom lands within U.S. jurisdiction but "out" beyond state-controlled waters (see discussion in chapter 2). In the oil-development business, "OCS activities" usually refer specifically to oil-and-gas extractive activities in OCS regions.
OCSLA	Outer Continental Shelf Lands Act of 1953, 43 U.S.C. 1331 et seq.
OCSLAA	Outer Continental Shelf Lands Act *Amendments* of 1978, P.L. 95–372, 43 U.S.C. 1801 et seq.
OMB	U.S. Office of Management and Budget
OPEC	Organization of Petroleum-Exporting Countries
OTA	U.S. Office of Technology Assessment
P.L.	"Public Law"—the number assigned to a bill as it is initially debated by Congress. After a bill is passed, its provisions are normally codified in terms of "U.S.C." numbers and implemented in terms of "C.F.R." numbers; see each acronym for further details.
U.S.C.	"United States Code"—the more-or-less topical codification of all the laws passed by Congress and still in effect.
USGS	U.S. Geological Survey, Department of Interior

LIST OF PHOTOGRAPHS

PREFACE

For decades, the largest official source of revenue for the federal government, outside of the Internal Revenue Service itself, has been the program to lease the offshore oil resources that lie beneath federal waters. In recent years, the program has become one of the government's biggest headaches as well. From Maine to California, and from Florida to Alaska, the government's efforts to expand offshore drilling have run into a virtual storm of protest. The protest is not universal—it is notably absent in the region where the offshore oil industry originated, along the shores of the central Gulf of Mexico. In other areas, however, the problem has become so pervasive that what was once intended to be a "national" program is increasingly becoming, in effect, a Gulf of Mexico program.

While the stormy reception has frequently been discussed, and just as frequently cussed, actual public attitudes toward OCS development have received little in the way of systematic study. The failure is in many ways a strange one; there is, after all, no law of nature that requires us to stop looking for solid information simply because questions of human behavior are involved. One of the results of the failure, however, has been that if we would like to be able to make well-informed decisions about the issue, we are faced with an unfortunate imbalance between assertions and evidence.

It is not at all difficult to find "experts" who will offer their views on the underlying causes of the storm, as well as their own forecasts and preferred "solutions." What is difficult to find is balanced analysis, backed up with solid evidence. As will become clear in the pages that follow, any number of the available explanations are simple and apparently plausible. In many cases, unfortunately, they also tend to show at least some signs of being self-serving, and in most cases, they prove on closer inspection to be wrong.

In this book, the two of us try to provide an antidote. Our goal has been to contribute the kind of analysis that seems to us to have been largely missing from the discussions to date—one that is intended to improve our understanding of the issues, rather than to improve the footing for one side or another, in the ongoing policy debates.

While the storm of protest has been widespread, there have been local and regional variations; the starting point for our analysis is the belief that, if we want to learn more about the underlying dynamics, these variations deserve closer attention. Our focus, more specifically, will be on the two regions that provide the sharpest contrast in terms of public reactions—southern Louisiana and northern California. Southern Louisiana has long provided the offshore oil industry with what may have been its most placid reception, while the coast of northern California has been the place, to date, where the fury of the storm has been most intense. For those who are so inclined, accordingly, this book can be read as a straightforward, comparative case study, one that seeks to bring greater light to a set of debates that are already characterized by more than enough heat.

At a somewhat broader level, the book is also intended to provide a balanced discussion of the most important of the underlying complexities, recognizing that the importance of these complexities has to do both with their implications for the offshore oil debate, in its own right, and with the fact that they can offer a potential source of insights about a relatively new but growing group of disputes. In these disputes, increasingly, we find that significant fractions of the citizenry become organized in opposition not just to industrial interests, but also to at least a major segment of the scientific and technical community, and to the government that, in a democratic system, is expected to represent the will of the governed. The battle over offshore drilling offers an opportunity to improve our understanding of the underlying reasons.

At the broadest level, the book is also an effort to contribute to ongoing discussions about a number of wider themes—themes about technological controversies, resource policy, the credibility of federal government agencies, and about relationships between society and the biophysical environment. These themes have already attracted the attention of a distinguished group of academic analysts; we have drawn heavily from the insights of these analysts, as will become clear, and we hope we have repaid the resultant debt by offering something of intellectual value in return. We share the belief of many of these analysts that the themes are serious ones, and we hope we have been faithful to their examples in terms of the seriousness of our search for insights. We have tried hard, however, not to follow the example of their language. We do not share the belief that the seriousness of the themes should be matched by the use of language that has the greatest gravity imaginable. Instead, we believe that, in writing, clarity is at least as much of a virtue as is gravity. Given that many of the people with an interest in the topics of this book will not be technically trained social scientists, we have tried to write in a language that closely resembles English.

One of the ways in which this will show up is that we have been sparing in the use of legal and bureaucratic jargon. We will refer to a number of the relevant laws, but only because it is sometimes necessary to be specific, and we will use the most common of the acronyms where they are helpful. For the most part, however, our discussion of legal technicalities has been banished to the appendix, where they can be consulted by those who wish to learn more, and avoided by those who do not.

We will also minimize our use of academic jargon. The effort to write in English, in fact, is simplified by the fact that the two of us have already published a number of technical articles based on the research summarized here. For those who wish to thrill to our treatment of serially correlated errors and our deft handling of Durbin-Watson statistics, we have provided references throughout that will provide access to the more technical articles. For this book, we have attempted to avoid the use of technical lingo as much as possible, to tell the story clearly, and to limit ourselves to just those tables and figures that appear to us to be most relevant to the broadest number of potential readers.

We will try to be equally brief in discussing our methods. In essence, we have relied on three main types of information. First, we have analyzed information from a variety of documentary sources. These include a quantitative content analysis of the comments on the draft Environmental Impact Statement that outlined potential plans for leasing oil off the coast of northern California. Quantitative analyses have also been done on other available data, which have been drawn not simply from statistical compilations, but also from less traditional sources, such as road maps and topographical charts. Additional background information has been obtained from existing histories, articles in the local and national media, and more technical materials, both from academic or "open" journals and from the agency's own reports.

Second, we have drawn on several dozen in-depth interviews, all conducted over the period from 1989 to 1992, with the majority of the interviews taking place in southern Louisiana and northern California during the summer of 1990. Our study was done with federal agency support—more on this in a moment—and this meant we were bound by little-known regulations from the U.S. Office of Management and Budget, which effectively preclude the use of survey research in most of the work that is done for federal agencies. The net result was that, for our interviewing, we were prevented from using one of the most basic techniques in the social sciences, the statistically representative survey. For most studies, this limitation is a major concern, and even in this case, it would have been helpful to be able to use a survey to examine some of the subtle nuances of local reactions. As will soon become clear, however, while some of the nuances could still benefit from a more detailed analysis, the overall contrast between northern California and southern Louisiana could scarcely be more clear-cut.

Third and finally, we have drawn on our own first-hand experiences and observations, more broadly. This means that, if only in the interest of fairness, we should provide readers with at least a bit of background information on the two of us, and on the origins of this study. Each of us has spent considerable time and energy in attempting to understand the development of the offshore oil controversy. Over the past decade or more, we have been able to watch, at reasonably close range, the increasing problems encountered by federal efforts to lease the offshore oil resources of the country. As a result of our efforts to examine the problems, each of us has been asked by the National Academy of Sciences/National Research Council to assist in evaluating the adequacy of the scientific basis for offshore development. We have been serving on resultant committees even as we have been writing this book. One of us has been serving on a committee that has had the responsibility to evaluate the overall adequacy of the scientific studies performed by the Minerals Management Service—the federal agency within the Department of Interior that bears responsibility for the offshore oil program—as well as the adequacy of the scientific data base for proposed lease sales off the coasts of California and Florida. The other has been serving on a committee with a similar charge but different geographic focus, evaluating the scientific basis for leasing off the northern slope of Alaska. As further indicators of the sources of our first-hand observations, one of us was also just finishing six years of service on the Scientific Advisory Committee for the Minerals Management Service at the time when the official data collection for this study was getting under way, and the other has also lived for more than a decade in Lafayette, Louisiana—the de facto headquarters for much of the exploration and development that has taken place in the Gulf—where he has been systematically studying the consequences.

     As for the present book, more specifically, it grows out of what may have been the first social science study, beyond the discipline of economics, to have been supported by the Minerals Management Service outside of the state of Alaska, although additional social science studies have since received agency backing. The decision to begin supporting social science research may have had something to do with the agency's recognition of the growing storm of protest, but it may also reflect a good-faith effort to respond to repeated recommendations from scientific advisory committees, as well as a growing awareness, within the agency, of the specific requirements of the law, the relevant portions of which are noted in the Appendix and in some of the later sections of this book.

     We appreciate the fact that the agency decided to support this study. Our appreciation is partly due to the fact that it was funding from the Minerals Management Service that made the study possible, but it is also partly due to the broader fact that, in deciding to move toward support for social science research, the agency is indeed showing progress toward a better ratio between evidence and assertion. We wish to stress, although it will probably become clear enough in the pages that follow, that while the Minerals Management

Service and the Department of Interior deserve credit for supporting this study, neither these nor any of the other agencies of the federal government deserve any of the responsibility for the discussions and conclusions in this book, all of which are exclusively our own.

The two of us truly do consider ourselves to be loyal friends of the agency, as well as of many specific individuals within it. Our belief, however, is that one of the duties of a friend is to point out even the kinds of facts that are uncomfortable, and that a mere acquaintance might prefer to leave unmentioned. Throughout this book, accordingly, we will attempt to provide the kind of balance that one would hope to have from a loyal friend—complimentary where possible, honestly and constructively critical where not. Our hope is that we have been able to do so in a way that will be helpful, both in dealing with the technological controversies in general, and in dealing with the troubled waters of offshore oil.

1

Dawn of Discord

Even before the first rays of sun managed to climb over the coastal hills of northern California, it was obvious that February 3, 1988, was going to be a gloriously beautiful day. The skies were so clear that the windshields were coated with frost, but the chill added an extra measure of freshness to the morning. The air had the crisp, clean smell of the sea; a full moon was setting into the Pacific; and the morning was serenely quiet.

Along most of that stretch of the coastline at that hour, the only sounds came from the surf and the seagulls, but in the picturesque coastal town of Mendocino, there were sounds of human voices, as well. The voices belonged to people who were not just out to enjoy the views of the ruggedly natural coastline. Dimly at first, and then more distinctly, the pre-dawn light revealed a group of protesters, complete with signs and baby carriages. They were bundled heavily against the morning chill, but otherwise high in spirits, talking excitedly with one another as they jogged, marched, and ambled northward along the coastal highway. By the time the first rays of sun brought warmth to the ground, and to the people, the group stretched out for more than a mile along the highway.

Their walk would have been warmer if they had waited for the sunlight, but few of them even considered that option. Their destination was the tiny town of Fort Bragg, about a dozen miles up the coast, where they planned to head for the local Eagles Hall. It offered the biggest room in town, with a capacity for several hundred people, but they were worried it would be full before they got there.

Like the Eagles Halls in most small towns, this one was often empty, and the event that was scheduled there that morning was the sort of thing not normally associated with standing-room-only excitement. A federal agency was scheduled to hold a hearing on a draft version of a technical report. The hearing was not even scheduled to start until 9:00 that morning. In this case, however, the technical report had to do with a proposal to sell offshore oil leases, and the betting was that anyone who arrived much after 7:30 would be too late to get in the door.

The betting proved to be right. By the time the slower of the marchers managed to make their way to the Eagles Hall, the security guards had

1

stopped letting anyone in, and the crowd outside had already swelled to more than a thousand. Estimates vary, but for much of the hearing that was to follow, just the crowd that was outside, waiting to get in, would swell to somewhere from 3,000 to 5,000 people. And things were no longer quiet.

It was not exactly as if the protesters on the outside found themselves with nothing to do. Anticipating the crowd, agency personnel had managed to hook up a loudspeaker, making it possible for at least one small segment of the crowd to hear the comments being made inside. A church down the street was broadcasting the event on closed-circuit television. The street had been cordoned off, and there was no real scarcity of interesting diversions for those who were unable to get in. Musicians played; officials made speeches; at least one politician's motor home dispensed coffee, donuts, and campaign literature, although the first two ran out before the third. A number of people simply enjoyed both the spectacle and the sunshine, with the weather warming up to its promise of a sensational day.

For those who were so inclined, it was also possible to become part of the spectacle, joining in with some of the ongoing demonstrations. In one of them, for example, organizers put together a "slick-in"—recruiting volunteers, providing them with the appropriate costumes (black plastic bags, of course), and even giving them some training in the rudiments of "oozing" and "sliming," before the event got underway. The demonstrators were obviously having such fun that the "oil slick" managed to recruit more volunteers as it proceeded, visibly growing as it oozed its way down the street. The onlookers enjoyed themselves, too, as did those who continued to make signs or speeches, or who hung politicians in effigy. A few members of the still-growing throng even expressed sympathy for the poor people who had to be cooped up inside on such a lovely day.

The several hundred souls inside the Eagles Hall may have been vaguely aware of the activities on the outside that they were missing, and all of them knew about the beautiful weather. Few of them, however, showed any inclination to leave, to enjoy the weather themselves, or to make room for the long, long lines of people who still hoped to get in. The activities outside, while lively enough in their own right, were little more than a side show. The main event was taking place inside.

Up at the front of the room, sitting at a long table, were a set of officials from the U.S. Department of Interior, from Washington, D.C. A local activist was to say later that she "felt sorry" for this group of people, in that they were in a kind of limbo—"high enough up in the agency to be here, but low enough that they had to sit there and listen to us." For the most part, they were also "high enough up in the agency" to have learned not to show much in the way of emotion when they were presiding over a public hearing, but to

Sign at Motel in Fort Bragg, California, 1988, exhorting people
to make their opinions known.

Protesters at the Fort Bragg hearing, the same day, making their opinions known.

the extent to which it was possible to tell anything from the looks on their faces, it appeared that they might have felt a little sorry for themselves too. In a way, that was too bad. Except for the fact that they "had to be there," they might even have enjoyed the proceedings. Their task, at least according to the official notices, was something that would scarcely have been expected to be one of the most memorable events in years—it was the "Hearing on the Draft Environmental Impact Statement for Lease Sale 91." In reality, what was playing out in front of them included great theater as well as a great outpouring of human emotion. Interspersed with the usual, technical assessments—a misplaced punctuation here, an erroneous resource estimate there, a missing evaluation of impacts in a given area of concern—came a series of comments that were neither typical nor technical. Hundreds of citizens had signed up, many of them weeks in advance, in hopes of letting the bureaucrats know just how strongly they felt about the issue of offshore oil drilling along their coast. It was clear, in fact, that at least some of the people in attendance did consider this to be "their" coast, although a larger number made the point that they considered the coastline to be a national treasure, with many of the latter group noting that they were speaking on behalf of a much wider public.

In their effort to "reach through to the people behind the facades," as one of the citizens later put it, and to make their case to the stone-faced bureaucrats at the front of the room, the people presenting testimony were also offering a bit more creativity than your average technician. Some of the voices thundered, others whispered, and still others choked with emotion. Guitarists performed songs that they had composed for the occasion; school children sang music and performed plays that they had written as well. At one point in the hearings, a man dressed as "Coyote" came up to the podium. He noted with approval the representation of his friend the buffalo on the Department of Interior seal, but then asked melodramatically, "*Where* has the buffalo gone? Are you *protecting* the buffalo? . . . Coyote's worried!" At another point, in the effort to demonstrate what a California-style earthquake could do to an offshore oil platform, a belly dancer, suspended horizontally by a set of accomplices, placed a plastic rig on her mid-section and began to gyrate. The oil rig performed its part of the drama impeccably, and the crowd roared its approval.

The crowd roared in other ways, as well. Well-prepared organizers had printed hundreds of cheap but effective placards, with black ink on 11" by 17" pieces of colored cardboard, each of which featured the word "NO," a diagonal slash through an oil rig in the middle of the "O," and a small black footprint. In addition to the placards, which were frequently brandished, the entire auditorium was decorated with signs of almost every size and description, save that none of them indicated much in the way of support for the proposed "lease sale." Most of the signs also used terminology and images that were far more colorful than the beleaguered agency representatives might have preferred.

While some of the best showmanship was to wait until later, the first handful of speakers gave some indication of what was to follow. The first speaker, the California Lieutenant Governor, left no doubt about his opposition, and the reaction of the crowd left little doubt that they agreed. The second speaker, the Chair of the Coastal Commission, spoke of his Commission's responsibility to protect the coast, and he made it clear that he would not take that responsibility lightly. The California Attorney General said not only that he was unalterably opposed to the federal proposals, but that he considered them to be illegal; he promised that if the sale were to go forward, he would take the agency to court, and that he would prevail. A member of the local County Board of Supervisors, noting that the Board had voted unanimously to oppose the sale, was more than willing to join in: "We will fight you page by page through this nefarious document." A Supervisor from Sonoma County underlined the point: "Welcome. You are surrounded. We ask your unconditional surrender." The idea of offshore drilling along this coast, he announced, was "as ridiculous as paving the Grand Canyon."

Yet that was only the beginning. The officials were to hear similar messages from representatives for both U.S. senators from the state, for twenty-three Members of Congress, and for a string of cities and counties, all along the coast. A social scientist in the audience, long accustomed to "keeping score" of support and opposition in hearings such as this, decided there was little need to continue once the first twenty speakers in a row voiced levels of opposition that ranged from polite to adamant, without a single speaker voicing even lukewarm support for the proposed leases. All the while, not content simply to roar their approval, the crowd cheered, screamed, and even chanted—"No! No! No! No!"—waving the signs that carried the same message, and stamping so emphatically, in rhythm with the chant, that the building literally shook. The people who were outside might have been enjoying themselves, but those who were inside were making a statement—and in a way, they were making history.

The hearing, once scheduled to "continue until 8:00 P.M., or until all testimony is received," went on into the wee hours of the next morning; the bureaucrats at the front table had just a few hours of sleep before the entire scene was recreated in new ways the next day. Again the next day, the proceedings stretched well past midnight before the agency representatives—and those who felt sorry for them—pleaded about prior commitments, about others who had reserved the hall, and about the fact that, after all, given the consistency of the testimony, they were getting the message.

So were people outside of the room. The then-current issue of *North Coast News* ("The Coast's Home-Owned Newspaper") had carried the headline, "Activists Hope For Invasion By Land," noting that local activists were "unabashedly hoping that a brief invasion by national media" would "help thwart an invasion by sea-borne oil rigs" later (Waataja 1988, 1). One of the earliest speakers, referring to the man who was then the U.S. Secretary of Interior, added that his hope was for the hearing to be remembered as "Hodel's last stand." As the hearing proceeded, there was evidence that the hopes were not to be entirely in vain. All of the major networks had their cameras there, complete with camera trucks and satellite dishes outside, as did a number of independent news outlets. All across the country that night and the next day, citizens and leaders alike were to be able to get at least some small inkling of the outpouring taking place. What they would see would easily justify the prediction on page one of that day's issue of the Santa Rosa *Press-Democrat*, that the hearing would be "the biggest political event on the Mendocino County's coast in anyone's memory; the kind of happening that people will talk about for years" (McKay 1988, 1).

By the second day of the proceedings, several hundred miles down the coast at a meeting of the same agency's Scientific Advisory Committee on offshore oil development, the social scientist who had been taking notes at the hearing reported to his colleagues that he was "still dazed" by the intensity of the reaction. He put that fact in context by noting that, far from being a stranger to hostile hearings, he was a specialist in technological controversies, someone who studied facilities that had all the popularity of nuclear waste repositories and toxic waste dumps. He had spent a total of nineteen hours in travel, just to be able to observe a half-dozen hours of the hearing, but he found the spectacle so striking that he said he would be "more than willing to do it again" (see also Botzum and Garner 1988).

Thousands of miles to the east, in Washington, D.C., the top officials of the agency were perhaps a bit dazed as well, although it is safe to say that they had little enthusiasm for going through such an experience again anytime soon. At least to date, they have not had to do so—but therein lies a story as well.

So intense were the feelings on the offshore oil issue that the California congressional delegation, reflecting the wishes of their constituents, had already succeeded in imposing congressional moratoria on offshore leasing anywhere along this stretch of coast, even before the Fort Bragg hearing. Despite brave protests from officials in the Department of Interior and the White House, additional moratoria were destined to follow. Within a year of the Fort Bragg hearing, in his first budget message to Congress, the freshly elected "environmental president," George Bush, was to announce that, in the interest of making decisions on the basis of facts and not just emotions, he would ask the National Academy of Sciences for an assessment of whether or not the government already knew enough to proceed with leasing. His announcement was to cover not just the coast of California, but also the coast of southern Florida, where protests had been at least slightly less graphic, but where the Republican governor of the state already had taken the federal government to court in the effort to stop the proposed lease sale there.

If he had known about the answer in advance, President Bush might not have asked the question. The National Academy report was to conclude that certain kinds of scientific evidence were in better shape in some regions than in others, but in none of the relevant regions was it possible to say that the agency already had enough information to be able to move forward with selling offshore. The report went on to say that a pair of bureaucratic distinctions treated as having considerable significance within the agency—arguments that a lease did not automatically confer a right to begin exploratory drilling, and that a company finding oil would not automatically be granted permission to start pumping or ''producing'' the oil—appeared in reality to have little practical importance. Technically speaking, the agency may have had the right to intervene at any point in the process, but no one could think of a single case where a company had leased a tract but been denied permission to begin exploratory drilling, or had found commercially valuable quantities of oil or gas without then receiving permission to move into production.

Almost exactly two and a half years after the Fort Bragg hearing, on June 26, 1990—the same date when major newspapers around the country were carrying the headline that the President had ''moved his lips'' on his ''no new taxes'' pledge—newspapers in northern California were featuring a different bit of news from the White House. President Bush, the long-time Texas oil man, declared a moratorium on any plans to develop offshore oil along the coasts of central and northern California, or of southern Florida, until beyond the year 2000—after his own term of office would have ended, even if he had been able to win re-election.

While politicians are often accused of stalling on topics they find to be too sensitive, this one was different: aside from the fact that the president's decision would mean that the oil in these two regions would be inaccessible to his friends in the oil industry, the program for leasing offshore oil reserves is one of the few in the entire federal government that makes money, rather than just spending it. Over the thirty-four years before the Fort Bragg hearing, the annual revenues for the federal government from the offshore leasing program had averaged almost $3 billion per year, making this program second only to the Internal Revenue Service as an official source of money for Uncle Sam (Louisiana Department of Natural Resources 1989).

Meanwhile, Back at the Gulf . . .

The morning of February 3, 1988, had also been a relatively quiet one in the Gulf of Mexico. There were no mountains here, but the setting of the same moon, and the rising of the same sun, were observed over the waters of the Gulf of Mexico by a fisherman who was several miles out to sea, off the coast of Louisiana. Even in the gray light before dawn, his position in the Gulf would have been easy to spot; just a few yards away, lit up with a kind of intense, industrial light that many people associate with an oil refinery, was something that obviously had not been put there by Mother Nature. As the sunlight stretched across the waters and the darkness dissolved, any of the people who were assembled at the Eagles Hall in Fort Bragg would have been able to tell that the man was fishing, literally, in the shadow of an offshore oil platform.

The fisherman was not there by chance alone, and he was anything but lost; through years of experience, he, and many others, had learned that the fishing was better next to the rigs. Other than thinking of the rig as a kind of handy, man-made reef in an otherwise silty sea, the fisherman thought nothing about it. He had no idea that so many thousand people, living along a different coast of the same United States, would have been so worried that just the same kind of structure would suddenly start to appear along their coastline. For him, the oil rig was as much a part of the setting as were the smell of the salt or the sound of the seagulls. Indeed, off the coast of Louisiana, it was difficult to get out of sight of an offshore production platform such as the one to which the fisherman was tied: by 1988 there were over three *thousand* platforms in federal waters, over three miles offshore, with many more in the state waters closer to shore.

About a month later, an actual lease sale was held in New Orleans for the central Gulf of Mexico. Representatives of the same arm of the Department of Interior offered for sale 33,580,616 acres of the Gulf bottom; 3,416,759 acres were actually leased. There was no need to arrive early; there was no crowd outside, waiting to get in, and no crowd inside to cheer or protest. Reaction to the draft Environmental Impact Statement for this sale (which was made public about a year in advance) can perhaps best be summarized in the only official response from the state of Louisiana, a letter from Governor Edwards to the regional director of Minerals Management Service (Minerals Management Service 1987, D.76):

Dear Mr. Pearcy:

Thank you for your letter of April 27, 1987, regarding LE-2.

I appreciate your taking the time to provide me with the Environmental Impact Statement for Gulf of Mexico oil and gas lease sale proposal [sic]. I am forwarding this information to Mr. Jim Porter, Secretary of the Department of Natural Resources, for his review and comments.

If I can be of assistance in any way, please let me know.

Kindest regards.

Sincerely,
EDWIN W. EDWARDS

There were no comments from Mr. Porter, who was soon to become the president of Mid-Continent Oil and Gas, one of the largest industry lobbying groups in the state.

THE PARADOX OF THE PROTEST

In case it is not already clear, these two scenes were separated by more than a few months in time and a few thousand miles in space. They are also united by a paradox. In certain coastal regions, most clearly typified by central and western Louisiana, offshore oil development has long been welcomed with open arms; in others, obviously including northern California, just a proposal for the same kind of development is almost enough to open armed warfare.

What explains the difference? A number of "obvious" explanations are available, but as will be noted below, while they tend to sound straightforward and plausible, they also tend to be wrong. If we are to understand the paradox, we need to examine the facts more closely. We need to learn more about the people, the regions, and the issues—and that is what this book is all about.

The remainder of the book is divided into six chapters. Chapter 2 provides an overview of the historical context within which the outer continental shelf (OCS) controversy has developed. Chapter 3 summarizes the results of a number of interviews that the two of us have conducted, comparing reactions in northern California with those in southern Louisiana. In the interest of providing an independent, quantitative assessment, this chapter also analyzes the responses to the Environmental Impact Statement (EIS) that was the focus of so much attention in Fort Bragg.

While there is a good deal that is important in the subtleties of the interview findings, the overall thrust will be that the underlying distributions of

Attitudes toward oil in Louisiana: Signs at "The Oil Center," Lafayette, Louisiana.

Attitudes toward oil in California: Signs at the Fort Bragg hearing, California.

opinion in the two focal regions does indeed appear to be just as polarized as might be expected on the basis of the vignettes in this chapter. In Chapter 4, accordingly, we shift from decribing the differences in attitudes to analyzing them, identifying the factors that appear to be responsible for the dramatic differences in viewpoints across the regions. As will be seen, the underlying factors are neither as simple as they appear to some, nor as mysterious as they appear to others; to achieve a reasonable level of understanding, however, it is important to combine the kinds of considerations that are often found only in very different academic disciplines, considering historical factors, the nature of the biophysical environment in each region, and the social realities of the human environment as well.

Chapter 5 begins to explore the factors that seem to have contributed to the failure of a reasonably well-funded Studies Program to identify the sources of many of these problems in the past. While the discussion of shortcomings is quite straightforward and frank, Chapter 6 is an attempt to balance the criticism with an equally straightforward set of suggestions, in the form of a framework for guiding the studies that still need to be done. In Chapter 7, in closing, we take a second look at some of the most common of the "common-sense" arguments about the opposition, finding that the arguments may have been wrong, but that they may also have been politically useful, at least in the short term. As we suggest, however, the past usefulness of those arguments may need to be re-examined; while helping the agency and the industry to deal with (or to avoid) a number of past political pressures, the arguments today may be contributing to the problem of longer-term gridlock. To understand the reasons why, it is helpful to start at the beginning.

2

Getting to Gridlock

One of the protesters at the Eagles Hall on that frosty February morning said that he and his fellow Californians already knew why they were in Fort Bragg that day—because they cared about the coast and wanted to protect it. What he wanted, he said, was more emphasis on why "they," the oil companies and the federal government, were in Fort Bragg as well. "If you read this EIS," he noted, "there's all this stuff about how they want a 'domestic energy strategy,' but their only domestic energy strategy is to drain America first. There's really only one reason the feds have got an eye on the California coast—they're thinking about Big Oil and Big Money."

As will become clear below, we believe the facts are not quite that simple. It is important to realize, however, that some of the factors leading to today's oil rigs off the coast of Louisiana, and to the controversy about potential rigs off the coast of California, were originally set in motion more than a century ago. It all started soon after a so-called "Colonel" E. L. Drake completed the first known producing oil well, near the banks of Oil Creek, in northwestern Pennsylvania, in August of 1859 (for the history of the Pennsylvania oil fields, see Miller 1959; Darrah 1972).

By today's standards, Drake's well was almost hopelessly modest, involving a wooden frame, a drill bit powered by human labor, a depth of just under seventy feet, and a production of just twenty-five barrels per day. The effort was ridiculed at the time as "Drake's Folly," "because of its apparent worthlessness" (Martin and Gelber 1978, 464), but this small well was to be the start of what Sampson (1975, ix) was later to call "the world's biggest and most critical industry." It was also to be the start of the first "bust" in petroleum prices: "The year after the first discovery, the price of oil was $20 a barrel; at the end of the next year it was ten cents a barrel, and sometimes a barrel of oil was literally cheaper than a barrel of water" (Sampson 1975, 21).

Within just six years of Drake's achievement, the picturesque-sounding town of Pithole, Pennsylvania, a few miles to the east of Oil Creek, became what may have been the first "energy boomtown." Today, Pithole exists as little more than a ghost town and a footnote; its entire boom-bust cycle lasted a mere 500 days. Once, however, it was "the most legendary of oil-rush cities. At its peak, in 1865, it had only 10,000 inhabitants. . . . It was the centre of oil communications: the Pithole post office was said to be the third busiest in the States'' (Sampson 1975, 19).

What is more likely is that Pithole's was the third-busiest post office in the single state of Pennsylvania, but even at that, there is little doubt that the level of activity was phenomenal. A forest of oil derricks sprouted up in an area of just a few hundred acres along Pithole Creek, and the underlying formation provided drillers with a brief bonanza of 3.5 million barrels of oil. Although the town never got around to constructing a sewerage system, it did have fifty-seven hotels, with some of them providing their patrons with delicacies such as fresh oysters, even though the nearest salt water was some 200 mountainous miles away. Pithole was also the location of "the world's first commercially successful crude oil pipeline'' (*Titusville Herald* 1962, 18), extending some six miles, and a railroad that connected the then-booming nerve center to other communities such as Oil City and Oleopolis (for a more detailed account, see Darrah 1972).

With the end of the Civil War in the spring of 1865, thousands of veterans joined the speculators who were drawn to the area by fantasies of fortunes. Yet very few of them "really made money in Pithole. The great majority came with little and left with less'' (Darrah 1972, 19). Drake himself was to die a pauper in 1880; the town of Pithole was to die even sooner. At what was once the site of the town, little remains today but a museum, a scattering of signs, and a few small pieces of the story that are preserved on the signs:

> The . . . future site of Pithole was a poor buckwheat farm in 1864 when the U.S. Petroleum Company leased a part of it to drill a well. When the well came in, the value of the Holmden farm jumped to $150,000 as it was sold to a pair of speculators. The farm was sold again for $1,300,000 and again for $2,000,000. Then the boom ended. The county paid $4.37 for the Holmden farm in 1878.

The reasons behind these spectacular changes in real estate values are not all that difficult to fathom:

> In January 1865, Pithole's first well came in, producing 650 barrels of oil a day. Later that month, two other wells came in, at 800 barrels each, and the rush to drill was on. . . . In fall of 1865, Pithole's wells produced over 900,000 barrels of oil, more than a third of the year's total production of 2.5 million barrels. . . . By January, 1866, Pithole's oil production dropped to 3,600 barrels a day, and by November to 2,000 barrels. A year later, Pithole's remaining few wells were producing almost no oil at all.

By 1870, within a dozen years of Drake's discovery, petroleum had been discovered and produced in a half-dozen states, including California—and

John D. Rockefeller and five other investors had set up the Standard Oil Company. The Rockefeller oil empire began with a refinery in Cleveland, about 100 miles to the west of Oil Creek, that had a capacity of 600 barrels per day, or about 4% of the refining capacity of the United States at the time. Within less than a decade, Standard and its associated firms controlled 90% of U.S. oil refining capacity (Clark 1987). In 1882, the company set up the Standard Oil Trust Agreement, the first national monopoly in the United States. Despite any number of attacks by reformers, the trust and its successor organizations have played important roles in U.S. petroleum politics ever since.

The Sherman Antitrust Act was passed by Congress and signed by President Harrison in 1890, due in part to the animosities toward the Standard Oil Trust that had been stirred up by investigative writers known as "muckrakers" (the best-known example of which is the later, book-length treatment by Tarbell 1904), but the act was vague and was only listlessly enforced in its early years. The Standard Oil Trust was supposedly dissolved by the Ohio courts in 1892, but Rockefeller was able to regroup by consolidating his holdings under Standard Oil of New Jersey. It was not until after Theodore Roosevelt was elected that the federal antitrust laws began to be enforced more vigorously; in 1911, still controlling 64% of U.S. refining capacity, the Rockefeller Trust was found by the Supreme Court to be in violation of the Sherman Antitrust Act. Standard Oil was divided into over thirty corporations (Clark 1987), but that was scarcely to be the end of "Big Oil."

Just a few years after the Supreme Court decision, while most industries (including coal mining) were experiencing heavy regulation during World War I, the oil industry managed to evade mandatory controls. Instead, the Oil Division of the U.S. Fuel Administration established a "committee system" to encourage "voluntary increases in production and efficiency," and the agency even argued successfully with Congress to increase oil depletion allowances in the War Revenues Act of 1918 (Laumann and Knoke 1987, 52). While arguments have since raged about whether the "oil depletion allowance" ought to be set at different *rates* (cf. Molotch 1970), U.S. tax policy ever since has continued to provide a subsidy to oil companies, effectively exempting oil companies from paying taxes on roughly a quarter of the income from producing oil wells (for further discussion, see Feagin 1990; Nash 1968).

Oil proved to have considerable military significance during the First World War. As of 1900, 90% of U.S. energy consumption involved coal, and coal remained the nation's primary fuel until the 1930s (Johnson 1979). The Allied victory in World War I, however, depended heavily on petroleum, which fueled military trucks, planes, and even ships. The effort to encourage "voluntary increases in production and efficiency," in other words, appears to have been reasonably successful. After the war, the oil companies were again unlike most of the other capitalistic interests in the country, being in no hurry to see the end of wartime controls.

One of the reasons may have been oil's potential for Pithole-style boom-bust cycles. Volatility was not much in evidence so long as just a few companies (or a federally approved system of voluntary cooperation) controlled production and prices, but at least in theory, the breakup of Standard Oil should have led to something more like a competitive free-for-all. The weakening of the Standard Oil grip should have been speeded up by a number of additional factors, including an earlier decision by the Texas courts to throw Standard Oil out of that state, followed by the discovery of massive Texas oil deposits starting in 1911, the same year as the Supreme Court decision that broke up the Standard Oil monopoly. In addition, postwar oil consumption soared, although new discoveries kept pace, with massive new oil fields being found in southern California and in Oklahoma during the 1920s. Overall, U.S. oil consumption grew thirtyfold between 1890 and 1925 (Clark 1987). The combined market share of Standard and its former affiliates dropped to 40 percent by the late 1920s (Nordhauser 1979), even though Standard Oil of New Jersey evaded the intent of the Texas legislation by secretly buying the Texas firm of Humble Oil (later to become Exxon, the world's largest oil company) in 1919.

At least for a time, the competition appeared to be truly large-scale; it even included international intrigues, such as secret deals with the then-new Communist leaders of the Soviet Union, which had been producing about 15 percent of the world's oil at the time of the Bolshevik revolution. The recriminations from the Russian intrigues, however, included a price war in India, a worldwide glut of oil supplies, and major economic headaches for the major oil companies, which saw a further reason for concluding that competition was contrary to their interests. By 1928–29, the major oil companies entered into the very kind of agreement forbidden by U.S. law, effectively creating an international cartel to control the production and price of oil on the world market. The Pact of Achnacarry,[1] as the agreement was known, would remain at least partly secret for roughly a quarter of a century (U.S. Federal Trade Commission 1952; Ghanem 1986).

As noted by Laumann and Knoke (1987), while federal government involvement with the oil industry during the first half of the twentieth century is often seen in terms of the enforcement (or failures of enforcement) of antitrust regulations, a second set of governmental interests had to do with the military significance of oil. We have already noted the military importance of oil during World War I, but by the time of World War II, the military connections were even more clearly in evidence. "Early in the war the possibility of oil shortages prompted the government to use funds from the Reconstruction Finance Corporation to pay for the construction of two major pipelines to the eastern seaboard" (Laumann and Knoke 1987, 58), and even the Secretary of Interior at the time, Harold Ickes, although generally remembered as being no great friend of the oil industry, arranged for that industry to be "secured immunity from a considerable amount of anti-trust regulation" (Laumann and Knoke 1987, 59).

It was not until after World War II that fuller information on the control of the world oil market by the cartel began to emerge. Although "national security" was evoked to attempt to suppress the information, the Federal Trade Commission revealed the details of the Pact of Achnacarry in 1952 (U.S. Federal Trade Commission 1952). The disclosures, coming on the heels of evidence that some of the major oil companies had cooperated with the Germans during World War II (Coleman 1989), led the Truman administration to initiate a grand jury investigation of the oil cartel, although the investigation was later undercut by the Eisenhower administration (Blair 1976; Kaufman 1978).

One result of the secret pact was that, for decades, the oil industry avoided any further cases of what might be called "the Pithole plummet." Cooperative control of the world market by the major oil companies remained in effect, with varying degrees of success, until the oil embargo of 1973–74. That the cooperation was more than tacit can be seen by the fact that antitrust regulations were specifically set aside a number of times during the 1950–73 period, allowing the major companies to negotiate as a group with various Mideastern countries, and after its inception, with the Organization of Petroleum Exporting Countries or OPEC (Ghanem 1986; Wilkins 1976). For much of the twentieth century, accordingly, oil was a carefully controlled and relatively stable commodity with a steady growth trend. During most of that time, federal policy was more notable for what it ignored in terms of antitrust violations, and supported through tax structure, than for what it regulated.

Oil in the Waters

After the end of World War II, however, the two areas of traditional federal government interest in oil industry affairs—military and antitrust concerns—were to be joined by a third. With the discovery of offshore oil, the development of the technology that was needed to tap it, and a series of decisions by the U.S. Supreme Court, the federal government came to have an income motive of its own.

In one sense, the offshore oil industry was half a century old by the end of World War II, and the industry was one that had started in California. Oil was first recovered from under the sea bottom in 1898, from piers into the Pacific Ocean near Summerland, California. Oil was also produced from wooden platforms in Cado Lake, Louisiana, in 1911. It was not until the 1920s, however, that wells were first drilled in coastal seas, at locations that were truly offshore. Those developments started and took shape primarily in the shallow, protected waters along the coast of the Gulf of Mexico.

The initial exploration and development entailed little more than the creative adaptation of existing, land-based drilling technology. Along the southern coast of Louisiana, the "land" is often almost as wet as the water, consisting largely of marshes, and exploration took place first in the marshes, only later moving offshore. The initial techniques were largely the same in either of these soggy environments, involving the construction of a platform on pilings that were driven deep enough to provide a solid footing, with drilling equipment then being brought to the site on barges and installed on the platform. From there, drilling proceeded much as it would have been done on land, except that transportation to and from the site was by boat, and additional costs were created by the need to build a platform for each exploratory well.

As the exploration moved into ever-deeper waters and encountered increased exposure to wave action, the non-recoverable cost of platform construction became considerable. Texaco's creation of the first mobile drilling barge, in 1933, pointed the way for a much larger offshore industry. The barge was designed to be towed to a site and sunk. Sitting on the shallow bottom, it could provide a stable base for drilling; once drilling was complete, the barge could be raised and moved to a new location. Texaco's original barge, the *Giliasso*, was designed for protected inland waters, but by the mid-1950s, the concept had been extended to marine drilling barges capable of working in the shallow Gulf. Drilling technology has advanced considerably over the ensuing decades, to the point where drilling has been done in water as deep as 11,700 feet, as part of the abortive "Mohole" project in the early 1960s, but the concept of mobility has remained a critical one for the development of offshore petroleum production (see Lankford 1971 for a more detailed history of the evolution of the technology; see also Stallings 1984; Brantly 1971).

In 1936, realizing the economic potential of offshore lands, the state of Louisiana created a State Mineral Board, directing the board to lease offshore tracts on a competitive basis. Success in the protected waters led to demand for access to prospects further offshore, which in turn led to conflicting jurisdictional claims by the affected states and the federal government. In 1945, the Truman Proclamations asserted the federal government's control and jurisdiction over offshore lands. In the absence of legal precedents to establish the federal title, however, Louisiana (and later Texas) proceeded to lease offshore oil lands following World War II (Mead et al. 1985). By 1946, drilling was being done six miles from land, southeast from Eugene Island, on a lease from the state of Louisiana (*Oil Weekly* Staff 1946). The work was done by Magnolia Petroleum Company—a Texas production com-

The Beach at Summerland, California, at the peak of "offshore" drilling, shortly after the turn of the century.
Courtesy of the Gledhill Library of the Santa Barbara Historical Society.

The view from Summerland, California, in 1992; note offshore rigs in distance.

pany that had been bought in 1925 by Standard Oil of New York, and that was later to become part of what we know today as Mobil.

The political and legal battles between the states and the federal government went on for years, but the outcome is one that will come as little surprise to anyone who is familiar with other such fights. The conflicting claims were ultimately settled by the U.S. Supreme Court between 1947 and 1950, in what are generally remembered as the Tidelands cases. This series of decisions (see for example *United States v. California* 332 U.S. 19 [1947]) established the legal rights of the federal government over all offshore lands (for further discussion, see Engler 1961, 86–95; Cicin-Sain and Knecht 1987).

In the wake of the Supreme Court decisions, the Tidelands controversy became a major issue in the 1952 presidential campaign. With the election of a Republican president who supported the states' claims, Dwight Eisenhower, Congress passed two landmark laws in 1953 that have continued to shape federal policy to the present. The first was the Submerged Lands Act of 1953, a compromise that assigned to states the title to offshore lands that were within three miles of the shoreline. Two exceptions involved Texas and the west coast of Florida, where the Supreme Court subsequently ruled that the states had held title to three marine leagues (approximately 9 miles), as sovereign nations, before they were admitted to the Union.

The second piece of legislation, which built on the first, focused on the so-called "Outer" Continental Shelf, or OCS—the sea-bottom lands "out" beyond the limits of state jurisdiction. This second law, the Outer Continental Shelf Lands Act (OCSLA) of 1953 (43 U.S.C. 1331 et seq.), authorized the Secretary of Interior to lease the OCS for mineral exploitation, through competitive bidding, and subsequently to administer the leases.[2] Initially, two agencies within the Department of Interior, the Bureau of Land Management (BLM) and the U.S. Geological Survey (USGS), shared responsibility for this task. The 1953 OCSLA initiated federal leasing for oil and gas (and salt and sulphur) on the Outer Continental Shelf, and the Act continued to serve as the major policy for the leasing effort between 1953 and 1978.

The first federal lease sale occurred in the Gulf in 1954, the year after the passage of OCSLA. A total of 394,721 acres were leased at this sale, which generated $116,378,476 in bonuses (Gould 1989, 50). A much smaller lease sale, for areas off the Texas coast, generated an additional $23 million in bonuses a month later, and a combined Louisiana-Texas lease in 1955 generated yet another $100 million. Additional monies began to flow into federal coffers from royalty payments, and the pattern was established.[3] From 1954 through 1969, there were twenty-one OCS lease sales, generating approximately $3.4 billion in bonuses alone, and the lease sales soon began to be recognized as a major source of revenue for the federal government.

At least in theory, the OCS program is a national one, but in reality, all but four of those twenty-one sales took place in the Gulf of Mexico. From 1959 onward, with only a few exceptions (1961, 1963, and 1965), lease sales have occurred at least annually in the Gulf; often they have been held two or three times a year. This level of leasing activity, in turn, generated some of the same kinds of agitated exploration seen much earlier around Pithole, Pennsylvania. In Louisiana, however, the boom lasted much longer, and the effects were felt across a much wider region, as will be discussed further below.

In other coastal regions, the level of leasing activity was much lower, but the levels of opposition, from the start, were much higher. The Department of Interior held its first lease sales on the central and northern Pacific Coast in 1963 and 1964; with this expansion to the Pacific Coast, the federal government began to meet resistance to its leasing program. In spite of local objections about the safety of drilling (see Molotch 1970), the federal government also leased tracts in southern California in 1966 and 1968. These southern California leases, however, soon led to a bureaucrat's nightmare.

On January 28, 1969, an oil-production well being drilled on Union Oil Company's Platform A blew out around its casing, spilling up to 3 million gallons of oil into the Santa Barbara Channel. The spill eventually affected over 150 miles of coastline, and with national media coverage, its political effects were still more widespread. The Santa Barbara blowout served as what Slovic (1987) calls a "signal" event. The spill has been credited, at least partially, with the passage of the National Environmental Policy Act (NEPA) in late 1969, the widespread impact of the first Earth Day in the following year, the eventual passage of the 1978 amendments to OCSLA, and the emergence of widespread, organized resistance to the OCS leasing program in most of the coastal states (cf. Cicin-Sain and Knecht 1982; Mead et al. 1985; Kaplan 1982). The spill also contributed greatly to the concerns over offshore drilling within the state of California. As part of the aftermath of the spill, a state-wide referendum in 1972 created a Coastal Commission, which has permitting authority over activities in the state coastal zone, and seaward to the limits of state jurisdiction. The Coastal Commission has since provided a unifying set of statewide regulations and a focus for organized efforts to protect the coastline (Kaplan 1982).

Following the spill, additional lease sales in the Pacific and initial sales in the Atlantic and Alaska were postponed; six years would pass before another OCS sale would be conducted outside of the Gulf of Mexico. In the Gulf, however, it was business as usual—only faster. From 1969 to the present, with the exceptions of 1971 and 1977 (which had one sale each), there have been at least two OCS lease sales in the Gulf every year. Between the 1969 Santa Barbara blowout and the passage of the 1978 amendments to the OCSLA, there were twenty-one more lease sales in the Gulf. Over the next two decades, the oil embargo and the growing federal deficit were to create additional pressures for offshore leasing, almost all of which would take place in the Gulf of Mexico.

Beyond the Blowout

During the early years of the 1970s, particularly after the 1973–74 oil embargo brought the "energy crisis" to the forefront of public awareness, two issues received the bulk of attention in the arena of petroleum politics—the Alaska pipeline and the oil embargo. In spite of increased public concern over environmental protection in the aftermath of the Santa Barbara blowout, the embargo increased concerns over the security of energy supplies. In a six-year saga, the battle over the pipeline led ultimately to Congressional decisions to settle long-standing claims of Alaska Natives, to overrule one of the nation's most important environmental laws (NEPA), and to open the oil companies' access to the estimated ten-billion-barrel reserve at Prudhoe Bay (see the further discussion in Gramling and Freudenburg 1992a).

The oil embargo also had effects on offshore leasing that require additional discussion. The Organization of Petroleum Exporting Countries had actually been founded well before it came to public awareness, having been born out of a meeting called by General Oassem of Iraq in September of

1960. The primary impetus for the initial meeting was provided by then-recent price cuts in Middle Eastern oil that had been initiated by the major multinational oil companies in 1959 and 1960. The reasons given for the cuts included a glut of tankers ordered while the Suez Canal was closed, and an earlier, less famous embargo—the 1959 *U.S.* embargo on *foreign* oil, which cut the demand for oil from what were soon to become the nations of OPEC (Ghanem 1986).

While OPEC had some limited success in raising the price of oil in the early 1970s, it was the oil embargo against the United States and the Netherlands, in retaliation for their support for Israel in the 1973 round of the Arab-Israeli war, that broke the control of the world oil market by the major multinationals. In a three-month period from October 1973 to January 1974, the price of Arabian light crude increased from $3.00 to $11.65 per barrel. Significantly, these increases were dictated by the oil-producing nations rather than the oil companies. The emergence of OPEC and the resulting 1973–74 oil embargo went a long way toward turning the multinationals into "suppliers of technology and marketing agents for OPEC oil" (Feagin 1985, 451). For all practical purposes, according to knowledgeable observers (see e.g. Ghanem 1986, 141–50), the embargo marked the end of the multinationals' control of the oil market.

In response to the embargo, President Nixon addressed a nationwide audience on January 23, 1974, to announce "Project Independence"—offering renewed support for other domestic energy measures such as the then-proposed trans-Alaska pipeline, but also directing the Secretary of Interior to increase dramatically the acreage leased on the OCS, to 10 million acres, starting in 1975. The plan involved leasing in every frontier area over the following four years, and represented more than a tripling of what the Department of Interior had originally planned to lease.

President Nixon's policies were not without critics. In particular, a staff report prepared by the National Ocean Policy Study for the U.S. Senate Committee on Commerce (Magnuson and Hollings 1975) raised four objections. First, the USGS estimates of the undiscovered recoverable oil resources on the OCS were far above all other estimates, including those by the oil industry, calling into question the assumption that the OCS could allow the country to achieve "energy independence." An alternative interpretation was that stepped-up leasing would simply deplete the country's reserves more rapidly, increasing national reliance on imported oil over the long term. Second, the report doubted whether the industry had the capacity (in terms of offshore rigs and support vessels) to explore the increased offerings as quickly as desired. Third, given the legal requirement that leases had to be explored within five years—plus the likely workings of the law of supply and demand—there was an obvious possibility that the increased offerings would result in a decline in revenues per acre for leases. Fourth and finally, the report argued, increased development could result in serious disruptions for coastal states. While the jury is still out on the initial objection, the other three objections were later proved to be squarely on target.

In spite of these problems, the Department of Interior moved toward increased lease sales on the OCS, and it did so in a way that indicated little concern about the objections of affected regions (cf. Wilson 1982, 74). Sales resumed off southern California in 1975 with the largest acreage offering to date (1,257,593 acres were offered; past sales in 1966 and 1968 had offered 1,995 and 540,609 acres, respectively). The first OCS sales off the East Coast and in Alaska occurred in 1976 and 1977, and there were three sales in the Gulf in 1975, offering the largest annual acreage to date, a total of 5,989,734 acres (Gould 1989).

Perhaps these massive offerings of public lands truly did reflect nothing more than a desire to "produce" or at least extract more oil domestically, in spite of environmentalists' objection that burning up finite domestic reserves at an increased rate provided at best a questionable "solution" to an energy crisis. Given the quantities of cash involved, however, it would be surprising indeed if federal policy were *not* to have responded to more than just the desire for energy independence. To date, the OCS leasing program has added more than $100 billion to the federal treasury (U.S. Minerals Management Service 1991). Such amounts would have an attractiveness to politicians under any circumstances; given the growing federal deficits, the sums appear to have become almost irresistible.

On the other hand, OCS leasing has proved to be significantly less popular among residents of affected regions, at least outside of the Gulf, than among federal accountants or devotees of "energy independence." The initial sales in Alaska and off the East Coast, and the resumption of sales in California, met with resistance from various interest groups and state governments (see the discussion by Kaplan 1982; Wilson 1982). Pressure also increased in Congress, leading eventually to the passage of the Outer Continental Shelf Lands Act Amendments of 1978, or OCSLAA. Congress was concerned that the OCS leasing and regulatory processes were essentially closed, involving just the Department of Interior and the petroleum industry. One of the purposes of the amendments was thus to open the decision-making process to a wider audience, thereby increasing public confidence in this federal activity (Legislative History, P.L. 95–372, p. 54).[4]

Under different circumstances, perhaps the reforms of 1978 truly could have led to increased public confidence. As matters turned out, however, they were followed within ten years by the revolt of 1988—with thousands of protesters descending on hapless federal officials, in the otherwise-peaceable town of Fort Bragg, in what scarcely seemed to be an outpouring of public confidence.

One of the things that happened in the interim, of course, was the election of Ronald Reagan, who ran in part on a platform of getting "burdensome" federal regulations "off the backs of the American people"—including, apparently, the petroleum people. Among its other actions, the Reagan administration squelched the Federal Trade Commission suit against Exxon and seven other major companies, which had dragged on since shortly after the embargo (Coleman 1989). Perhaps President Reagan's key action with respect to offshore oil development, however, came in the form of one of his most controversial appointees. The man he picked to head the Department of the Interior, or the very agency charged with administering the key provisions of OCSLAA, was none other than James Watt.

By 1982, within about a year of having become the Secretary of Interior, James Watt had reorganized the department's offshore-oil functions. Up until then, as noted above, the functions had been split between the BLM and the USGS; under Watt, they were combined within a single agency, one with a name that left little doubt about its primary mission—the Minerals Management Service, or MMS.

By 1983—no mean feat, given that leases are scheduled years in advance—Watt had succeeded in implementing another change, going to "area-wide" leasing on the OCS. This procedure opened entire "areas" for leasing (e.g., the entire unleased portion of the Central Gulf of Mexico) rather than offering leases only on designated tracts. As might be expected, this new approach greatly increased the acreage offered for lease: even the largest lease sale in the Gulf prior to area-wide leasing (in the "energy independence" days of 1975) had offered only 2,870,344 acres. In contrast, the very first area-wide sale in the Gulf offered 37,867,762 acres—an increase of more than 1,200 percent, even when compared against what previously had been the largest single lease sale in history.

Watt's successes, however, appear to have come at some cost. The first form of cost was economic: as predicted by the Senate committee report (Magnuson and Hollings 1975), the increased availability of acreage appears to have been at least partially responsible for a decline in the bonuses per acre. The first area-wide lease in the Gulf in 1983 resulted in an average bonus of $1,090 per acre, but by 1988 the average bonus had declined to $89 per acre (Gould 1989)—more than a 90 percent drop in bonus prices in just five years, or one of the most dramatic drops since the Pithole plummet. The second form of cost was political: area-wide leasing increased opposition to the OCS program, resulting in renewed political pressure at both the state and federal levels (cf. Wilson 1982), leading to the kinds of political and legal challenges noted in the introductory chapter, ranging from lawsuits to congressional moratoria. By the 1990s, the Minerals Management Service, whose fundamental goal is the leasing of OCS lands, effectively found itself denied access to many of those very lands. The third form of cost, which Magnuson and Hollings (1975) predicted, and which has received the least amount of attention to date, has to do with the socioeconomic impacts experienced in coastal communities and regions—a topic that deserves a section in its own right.

The Other Gulf

Ever since the 1973–74 OPEC embargo, and certainly since "the Gulf War" of 1991, many references to "the Gulf" in oil policy discussions have had to do with the *Persian* Gulf. The references are not altogether surprising, given that the majority of the world's proven oil reserves lie in this area. In the United States, however, is another oil-rich region, one that borders the Gulf of Mexico. Our references to "the Gulf" will have to do with the U.S. version, the Gulf of Mexico—but even in needing to make that point clear, we are perhaps illustrating the point that the human and environmental consequences of offshore oil activity in this region have often been overlooked.

Given the lack of local opposition, the high rate of *actual* leasing in the Gulf has often received much less attention than has the mere *potential* for leasing in most coastal areas. In general, while the oil experiences of the Gulf of Mexico region have been the focus of a few articles in the mass media, including occasional articles in periodicals such as the *National Enquirer* (Mullins 1981), attention has generally been limited to a number of more prosaic analyses in academic journals (see, e.g., Gramling and Brabant 1986; Gramling and Freudenburg 1990; Freudenburg and Gramling 1992).

Coastal Louisiana, however, is the region where the offshore oil industry was literally invented and developed. The vast majority of all the petroleum products from the U.S. outer continental shelf—well over 90% of the oil and approximately 99% of the gas—have been produced in the Gulf (Gould 1989, 99). Most of the Gulf's production, in turn—97.6% of the oil and 88.1% of the gas—has been within the central region of the Gulf, and within the central region, the overwhelming majority of production has been from waters adjacent to or supported from Louisiana (Gould 1989, 98). By all indications, current trends will continue into the foreseeable future: the majority of proven reserves on the OCS are in the central Gulf region, and even as of the most recent year for which data were available at the time of this writing (1991), Texas and the Gulf of Mexico OCS accounted for 61% of new crude oil discoveries (U.S. Energy Information Administration 1992). Just as importantly, in terms of practical consequences, the central and western Gulf regions are also the locations where MMS has encountered the least resistance to OCS activity.

From the 1950s through the early 1980s, local effects of OCS development produced three main types of impacts on coastal Louisiana. The first set of effects resulted from the leasing schedule itself. Leases expire, and bonuses are lost, unless production or at least exploratory drilling is under way in five years (or ten years in frontier areas; see 43 U.S.C. 1335).[5] Leasing thus generates high levels of inherently short-term offshore activity, including pressure for equipment (rigs, boats, etc.) and for what practitioners call "infrastructure" (the systems that underlie or support development, including everything from roads and harbor facilities to schools and social services). Given that the time period for initial activity on a lease is constant, and limited, an increase in the level of leasing translates quickly into an increase in the level of onshore as well as offshore activity.

These pressures are exacerbated by the fact that the most potent causes of growth-related social and economic impacts—the highest levels of employment, the greatest needs for secondary and tertiary support, and the most intensive requirements for capital—are all found during the phases of OCS activity that involve exploration (seismic and exploratory drilling) and development (construction of platforms, development drilling, completion, laying of pipeline, etc.). Once the platforms are in place and the oil is being extracted, the remaining activities are relatively routine. Platforms must be inhabited or monitored and maintained, wells must be cleaned out, and ultimately, once production stops, the platforms must be removed, but these activities require relatively few workers or new investments, particularly in comparison with the activities unleashed in response to leasing.

A second set of effects came about as a result of the price increases. As noted above, the OPEC embargo led to dramatic increases in the price of Arabian light crude, from $3.00 to $11.65 per barrel, just from October 1973 to January 1974, but that was scarcely the end of it. The price was ultimately to increase to $31.77 per barrel by 1981—an overall jump of more than 700% in seven years. The rising prices sparked an explosion of offshore activities, and of the demands on the coastal communities that supported the offshore industry. The economy of the central Gulf of Mexico region felt the shock waves of the explosion (Manuel 1984, 1985; Gramling and Brabant 1986), experiencing tremendous growth in the construction of offshore rigs and related activities.

Third, the "related activities" were not limited simply to those taking place along the Gulf of Mexico. While one common problem in the extraction-dependent communities of the later twentieth century is the difficulty of developing "linkages"—local economic benefits from selling supplies to extractive activities, for example, or from processing the raw

Louisiana coastal land use: Pipe yards at Morgan City, Louisiana, 1992.

Louisiana coastal land use: Offshore fabrication yards at Morgan City Louisiana, 1986, near the height of offshore activity.

materials before sending them out of the region (see e.g. Bunker 1984; Lovejoy and Krannich 1982)—coastal Louisiana achieved a level of "linkages" that would be the envy of most extraction-dependent regions. To drive through Morgan City, Louisiana, during the height of the boom, would have been to see a dizzying array of oil-supply firms—providing helicopters, drilling muds, seismic equipment, engineering services, clean-up crews, and dozens of other "input linkages." Louisiana also has one of the nation's biggest concentrations of oil refineries and petrochemical processing facilities, providing classic examples of "output linkages." Better still, in terms of the kinds of linkages that economic-dependent practitioners often only wish they could attract (cf. Feagin 1990), the state is also home to a number of regional and even national headquarters for oil-related firms. So extensive are the related developments, in fact, that even in the two Louisiana parishes (i.e., counties) having the heaviest dependence on oil extraction, *direct* employment in extraction never came to be as much as 15% of the total employment, even at the height of the boom (Gramling and Freudenburg 1990).

Instead, the development of Louisiana's linkages to the offshore industry eventually reached the point where the export of offshore oil technology became a major element of the coastal Louisiana economy. Even when offshore exploration was moving into "frontier" regions, including the North Sea and the North Slope of Alaska, much of the work was supported from the Gulf of Mexico. Because of the "concentrated" work scheduling associated with offshore work—where a worker can be flown to a remote rig for twenty-one consecutive 12-hour days of work, for example, followed by twenty-one days off—it is possible for individuals to live considerable distances from where they regularly work (cf. Gramling 1989). The 21/21 schedule, for example, requires only eight or nine "commutes" to the workplace each *year*, meaning that many of the offshore jobs in other parts of the world could be filled by skilled offshore workers from the Gulf of Mexico. In addition, many of the products purchased for offshore exploration and development are inherently mobile, meaning that they can be constructed in any coastal region in the world. The first platforms in the Santa Barbara Channel, for example, were built not in California but in Louisiana; of the five production platforms in place in the Pacific OCS prior to 1975, two were constructed in Morgan City, Louisiana (Gould 1989, 89). The net result was that linkages with support sectors, including both capital investment and labor, grew not only in response to continued activity in the Gulf, but also in response to offshore development in the rest of the world, throughout the 1970s.

By 1981, the northern Gulf of Mexico had become the most developed offshore area on the planet. The U.S. Minerals Management Service (1988) estimated the OCS program to have generated 190,000 jobs by the mid-1980s, with an annual payroll of over $4 billion; most of that economic impact was felt in the Gulf. The 13,000 + production wells drilled on the outer continental shelf alone (Gould 1989) yielded 9–12% of the crude oil and 10–20% of the natural gas produced in the United States from 1973–81 (Manuel 1984). Secondary and tertiary support sectors of the economy kept pace with the swelling primary sector, leading to explosive growth in employment and population (see Gramling and Brabant 1986; Gramling and Freudenburg 1990).

And then came the bust. Given that some of the same factors affecting the Gulf, such as the soaring prices of oil, were also at work in other parts of the world, production rose worldwide. As is the case for most commodities, however, the demand for oil is to some degree what economists call "elastic," meaning that the demand tends to go down as the price goes up. Much of the response in demand, however, is delayed, often taking place through new investments that are put into place only gradually. Being much less dramatic than the 700% increase in prices or the long lines at gas stations, the conservation-related changes were easy to overlook, but with the rising prices of crude oil and refined products, consumers gradually reduced consumption, primarily through quiet steps such as installing attic insulation and buying more fuel-efficient automobiles.

By 1981—the very year when James Watt began his drive toward more aggressive leasing of offshore lands—petroleum energy consumption in the United States had actually fallen below the levels reached before the 1973–74 embargo (*Oil and Gas Journal Data Book* 1988). Given the memories of the shortages and the soaring prices of the previous seven years, decreases in demand did not show up at first in the price of oil. Prices fluctuated but stayed relatively strong for several years, remaining at $24.51 per barrel as late as December of 1985. By June of 1986, however, the price had fallen to $9.39 (*Oil and Gas Journal Data Book* 1988, 101). Once the global prices began their steep dive, the local consequences were quick to follow: unemployment rates in the oil-dependent regions of coastal Louisiana, which had averaged in the 4–5% range for the previous decade, rose past 20% in many areas of the oil patch by the end of 1986.

The collapse of prices, and of employment, was greeted first with astonishment, then with anguish. Many of the region's younger residents had never known a time when coastal Louisiana was *not* booming, and at least at first, there seemed to be reason to hope that the sudden plunge would be a short-term aberration, to be followed just as quickly by a return to the only kind of reality they had ever known. If there are any residents who still harbor such hopes, they have already had a very long wait. As if to underscore the fact that Louisiana's energy-related bust was not a small or short-term matter, the January 1992 issue of *New Orleans Magazine* noted that, of the area's twenty-five savings and loans that had managed to survive for five years after the disastrous price drop that followed December 1985, only fourteen "remained untouched by regulators and met minimum federal capital standards" as of the period ending six months later (Finn 1992, 38).

Throughout the history of leasing on the OCS, Louisiana has officially presented a pro-development stand, but by 1991, there were initial signs of change. Realizing that over $100 billion had been generated for federal coffers by OCS leasing, and finding it increasingly difficult to maintain the physical infrastructure necessary to support OCS activities, the state initiated discussions with Secretary of Interior Lujan concerning revenue sharing. The discussions with Lujan, and later with MMS director Williamson, went nowhere. In August of 1991, for the first time in the history of OCS leasing, a frustrated Louisiana governor, Buddy Roemer, went to federal court in an attempt to block a lease sale in the Gulf. One of the state's major contentions was that the federal government had failed to address the social and economic impacts of the OCS leasing program in the Gulf. The state lost on a technicality, but by the end of 1991, the Department of Interior had presented a revenue-sharing bill to Congress. As of this writing in 1992, however, Congress had yet to act on the bill, and with the return to office of former governor Edwards, there has been an evident backing down from the state's once-tough stand.

The Local Effects

Not all of coastal Louisiana has been affected in the same way by oil development. Many of the regional headquarters buildings are located in New Orleans, a few blocks away from the old "French Quarter," along Poydras Street, but the real centers of activity are closer to the middle of the oil region. The clearest examples of all can be found in two counties, or "parishes," most heavily involved in offshore development—St. Mary and Lafayette parishes. The differing experiences in these two parishes help to illustrate the range of results of the boom-bust cycle, but they need to be examined one at a time.

St. Mary parish has an ideal location for offshore exploration and development activities, being situated where the Atchafalaya River cuts through the Teche Ridge (the natural levee of Bayou Teche) and intersects the Gulf Intracoastal Waterway. The Atchafalaya provides the only deep-water access to the Gulf between the Mississippi and Calcasieu rivers, a stretch of coastal marshes some 200 nautical miles wide; the Intracoastal Waterway provides east-west water transportation, and the Teche Ridge provides solid land in an area pinched between overflow swamps of the Atchafalaya Basin to the north and the coastal marshes to the south.

Before the onset of offshore oil development, local settlements, particularly Morgan City and Berwick, had responded to the potential of the local environment, developing an economy based on resources from the Atchafalaya Basin (fish, crawfish, cypress lumber, etc.) and the coastal marsh (oysters, shrimp, furs, etc.—see Comeaux 1972; Gramling 1980). Oil production led to a swift and dramatic transformation. Morgan City, the self-proclaimed "shrimp capital of the world" in the 1950s, had no resident shrimp fleet and no operating shrimp-processing plants by the 1980s. Employment in mining (the category that includes oil production) increased by 438% in St. Mary parish between 1940 and 1970, compared to 32% for the United States as a whole during the same period. Employment in transportation, an important "linkage" sector for offshore energy, increased by 412% (Manuel 1980).

As might be expected from the broader literature on the human impacts of rapid, energy-related growth (for a summary, see Weber and Howell 1982), the area exhibited many classic boomtown characteristics by the mid-1970s (Gramling and Brabant 1986). Housing was critically short, exacerbated by the limited availability of solid land. Entire communities were created; Bayou Vista, laid out in the 1950s in former sugarcane fields, was an incorporated community of over 5,000 by 1970 (U.S. Department of Commerce 1970).

The growth put strains on the oft-cited litany of facilities and services such as utility and sewage systems (Durio and Dupuis 1980), roads (Stallings and Reilly 1980), recreation facilities (Reilly 1980), medical facilities (Gramling and Joubert 1977), and especially education. Public school registration in the parish increased from 6,633 to 16,259, or 115%, in the two decades between 1950 and 1970 (Gramling and Brabant 1984; cf. Freudenburg 1982, 1984). Public utilities and services (water supplies, sewage treatment, utilities, recreational and medical facilities, police and fire departments, etc.) remained behind the population growth throughout the 1960s and 1970s; given that demand increased more rapidly than did the tax bases, there was little way for the supply of these services to catch up.

The 1981–86 downturn in oil prices brought a different kind of change. Before that time, the unemployment rate for St. Mary parish had remained low and stable for decades, standing at 4.6% for the year of 1980, but the rate shot up to 24.5% by 1986 (Louisiana Department of Labor 1970–87). Employment has since remained well below 1981 levels, leading to increased out-migration, business failures, and general deterioration of visible community infrastructure.

Lafayette parish has also experienced considerable impacts from the oil and gas industry, but the specific influences have been different ones. The parish seat and only city, Lafayette, developed as a railroad town and later functioned as a transportation center for nearby farming communities (cf. Gramling 1983). Lafayette parish also differs from St. Mary parish in that growth from the offshore oil boom came from administrative activities, rather than from drilling, transportation, and construction. Already a growing wholesale and retail center, Lafayette could attract oil-industry professionals by offering the kinds of urban amenities that the small communities in St. Mary parish could not provide. Lafayette became a headquarters location for many corporations, and its growth during the late 1970s and early 1980s also approached "boom" levels. Overall, Lafayette parish grew from 43,941 to 150,017 persons between 1940 and 1980, while the city of Lafayette increased from 19,210 to 81,861 (U.S. Department of Commerce 1940, 1980), increases of 241% for the parish and 327% for the city, as compared to population increases of 78% for Louisiana and 72% for the United States as a whole over the same four decades. Lafayette received national media coverage during this period, including hyperbolic attention in the *National Enquirer* (Mullins 1981), being reported (erroneously) as a place where two out of every thirty residents were millionaires, teenagers sported diamond-studded Cartier watches, and people hopped on Lear jets for lunch hundreds of miles away (see also Edmunds 1983).

As a "headquarters city," Lafayette would normally be expected to have been less susceptible to economic swings than would extractive communities such as Morgan City. In addition, although the growth in Lafayette was rapid, the area had a larger initial population and was less isolated than the typical western boomtowns that had brought the effects of rapid growth to the attention of social scientists (see Freudenburg 1982, 1986a, 1986b); many of the impacts traditionally associated with rapid growth thus were minimized or avoided (Gramling and Brabant 1986). Still, like most of coastal Louisiana, Lafayette experienced at least some of the stresses and strains that were associated with rapid, energy-related growth in other regions of the United States.

The Lafayette region also experienced the "bust," although the timing and magnitude were different. Growth-related problems began to moderate with the downturn in crude oil prices in 1982. With the crash in crude oil prices in 1986, unemployment in Lafayette parish rose, but the level reached "only" 9.6% in 1986—double the 4.8% rate of 1982, but far lower than the 20%-plus levels seen in St. Mary parish. Total employment remains below 1982 levels; property values have dropped dramatically, business failures are common, and out-migration is high. By the time this study's interviews were conducted in mid-1990, the unemployment rates had declined significantly, but the parking lots of shopping centers still were relatively empty—a fact that did not go unnoticed in California, half of a continent away.

3

The Citizens Speak

Late in 1987, less than a year before the showdown at Fort Bragg, a high-ranking MMS official told one of us that, while he had some frustrations about the views of the California general public toward offshore oil leasing, the frustration was mainly that "the truth" was not getting out. The problem, he said, was that the agency was only hearing from "the activists who make a lot of noise," but he saw them as a small and unrepresentative slice of the population. The "silent majority" in northern California, he continued, were actually in favor of leasing.

It is conceivable that the official in question actually believed his own claim, at least at the time when he made it. As will be seen, however, "the truth" actually stands in stark contrast to his claims. Rather than being found among a "silent majority" of northern California residents, support for offshore oil leasing appears to be limited to a small but relatively outspoken minority.

While politicians, of course, will normally maintain that they are already experts about "people," social scientists prefer to gather their data in more systematic ways. Often, as is the case for the present study, those more systematic ways start with the radical notion that, if we want to understand why people behave as they do, then we would do well to start by listening to, and learning from, what the people themselves have to say.

In the study reported in this book, we have tried to do just that. As noted in the preface, the study involved three main sources of information—documentary materials, the first-hand experiences that the two of us have accumulated over the past decade or more, and a series of in-depth interviews that we conducted in 1989–92, the majority of which took place during the summer of 1990. Given that the people who spoke with us in the interviews are the ones who, in many ways, know their regions best—they are the people who live there—we will summarize the results of the interviews first, bringing in the other forms of data later.

The interviews focused on at least two communities in each region—Lafayette and Morgan City in Louisiana, and Eureka/Arcata and Fort Bragg/Mendocino in California. Given the exploratory nature of this study, we used what social scientists would call a snowball-sampling approach. Persons who were known to be active and/or influential in OCS issues were among the first to be contacted; these people, in turn, were asked to identify others. Given that such persons are key sources of information for studies such as this one, they have come to be known, in the lingo of the trade, as "key informants." The people so identified included community political leaders, local business leaders and environmentalists, newspaper editors, fishermen, and others, including "common citizens" holding no particular positions of leadership or authority. The sampling in each community continued until we had heard from representatives of a full range of local viewpoints and the interviews revealed a clear pattern of redundancy, with subsequent interviews providing little more than a repetition of earlier themes (cf. Glaser and Strauss 1967).

These people were asked to share insights and views, to give their explanations for the local perceptions of the risks of offshore development (their own perceptions, as well as those of other residents), and to discuss the ways in which those perceptions had evolved. The sampling procedure for the interviews was not random, and the interviews should thus not be used as the basis for statistical analysis. Every effort was made, however, to develop a representative sampling pool, and as we interviewed people, we asked them to identify others who disagreed with them (and if possible, also persons who were relatively neutral), as well as those with whom they were in general agreement.

With the exception of one interview with a political leader in California who did not wish to be recorded, all of the in-depth interviews were tape-recorded and later transcribed. We took handwritten notes during all of the interviews, including the interview with that Californian politician. In the interest of encouraging frank discussion, and in accordance with customary procedures, all of these people were promised that their names would be kept confidential, although specific comments would be quoted without the use of names or other information that would reveal the identity of specific individuals.

Aside from limited editing for clarity and readability, such as the removal of the "uh" comments that we all use more than we realize until we see our comments transcribed, the comments below will be reported verbatim. Ellipses [. . .] will be used to indicate where words have been omitted to save space or improve the flow of remarks, and where the two of us have added comments or clarifications, these will be indicated [by the use of square brackets].

In the two sections that follow, we will turn first to the interviews from Louisiana, and then to those from California. For each region, we will summarize the remarks in terms of three categories of concerns—about economic prospects, local ways of life, and the biophysical environment. In the case of California, we will add a fourth category to deal with a topic that came up repeatedly there, although not in Louisiana, namely the credibility of the oil industry and the federal government.

Even a first-time visitor can easily see signs of the long-term influence of offshore oil exploration in Lafayette and Morgan City. Someone who moves to town can buy a house from Oil City Realty, go shopping at the Oil Center Shopping Mall, and go to work at an office in the Oil Center. Until the bust came along, that same person could have visited the Oil Museum on the highway between the two cities, and even today, to drive between the two is to see not just a great deal of real oil technology—pipes, platforms, jack-up rigs, crew boats, heavy trucks and helicopters—but even such nice touches as an office building that has been built in the size and shape of an actual offshore drilling platform.

This is not the kind of country, in short, that would seem likely to be the home of substantial numbers of people who are unalterably opposed to oil development. As the interviews revealed, it isn't. Still, given that unanimity is an unusual phenomenon in studies of people's attitudes—after all, a political election is considered a "landslide" if one candidate gets just 60 percent of the votes—we were more than a little nervous when our initial interviews revealed *no* evidence of outright opposition.

Just to be on the safe side, accordingly, we soon began to ask not just whether the persons interviewed, themselves, had any reservations about the continued extraction of offshore oil, but whether they *knew* of anyone who actually opposed it. Several names were suggested, and we did find people who expressed specific reservations, as will become clear from the comments below, but we found no outright opposition. What we did find will be summarized here in terms of the three sets of concerns just noted—about economics, ways of life, and the biophysical environment.

Concerns about Economic Prospects

Boom-Bust Experiences. Clearly, the most pronounced of the Louisiana interview findings had to do with the traumatic nature of the region's roller-coaster ride through the boom-and-bust cycle. The topic came up repeatedly, and often graphically, in our interviews. One planner summarized the boom days as follows:

> Morgan City's in our district, and I know that Morgan City, well up 'til recent times, this was really the area that attracted the population because of job development. I mean you could just, transients could come in, get a job offshore and make their money, come back in and spend it all on their seven

days off.[6] I think that is one community that is a very good example of, you might say, the beneficial aspect, and also the risk involved. Because at one time Morgan City had the highest crime rate in almost any area in the U.S. It just attracted everybody. I remember Mayor Brownell . . . he was on our board, he'd come over here looking for federal funding to build jails. I mean they had problems. . . .

People, they were coming in seven days on, seven days off, and when they'd come back in, they were in Morgan City. You had bars, you had fights, you had so much—crime was rampant in Morgan City. And the jobs were there. . . . Certainly the city had revenues, they prospered while the boom was going on, but here's a little town of about 16,000 . . . and it was just bottleneck traffic. You had real difficulty getting through Morgan City because of the activities going on. But now you go there, you see that it's quite different—a lot of business has gone. . . . [so have] businesses that serve the industry, the restaurants they'd had . . . [and] all these other offices that serve the population over there.

But I think that it was a very good example of what sometimes the oil industry attracts—the people that can't get jobs anywhere else. And when you need the workers, you'll just hire anyone. I remember reading accounts of some of the people that they found, some of the ten most wanted in America. They could look for them in Morgan City, for a while, and find a couple of them anyway. It was a lot of people.

The problems related to the bust were proving to be quite different. Various residents noted the problems of personal, family, and community stresses that had resulted from the sudden downturn in economic fortunes; others noted that the bust had created less obvious impacts as well. As one observer noted, for example, the area has found itself losing not just dollars, but talents:

A bank building [downtown] is going to lose the Texaco office it's had since the bank was built, and they're going to have to find somebody else to get in there so they can continue to pay for their building. . . . [But] the worst part about this is that these people were bona fide oil company office people; these people I would say were making an average of $80,000 per year, at least. And they were therefore white-collar people, they were therefore the people that you could call on to chair recreation committees, planning and zoning committees, help you with a church function or whatever. They were very capable, educated people, who knew what was going on, knew the other right people in town to contact to get this done or that done. And they're going, they're leaving. And Mobil Oil did the same thing. Now they are leaving, Texaco and Mobil both, are leaving the blue-collar and the labor behind. We will I guess eventually have to accept the fact that we are going to, for the interim anyway, be a warehouse-type district. Any of the pipes, any of the mud, any of the crew it takes to make all this stuff come together—they're going to be here. But they're not going to be, unless things change dramatically—I don't mean this in a derogatory fashion, but they're not going to be the JayCee presidents, not going to be Chamber or Board members, and the people we've become accustomed to having help us run the community. We're losing a top layer. It will be hard to replace.

While the boom times had clearly not been without their problems, those problems seemed quite mild in comparison with the human costs of the bad economic times that followed. Even though the region had been in the grip of the bust for roughly half a decade by the time of the interviews, the memories of what Gulliford (1989) calls the "euphoria" of the boom days were still vivid. As one person recalled, "During that period of time . . . it was a little different scenario: Money was rolling, people were spending, we were all *flying high* [laughs]—buying Mercedes. . . . We thought, oh, the world would never end." The benefits most often noted had to do with the direct economic advantages, although as this woman emphasized, the region also enjoyed many of the "linked" or indirect economic benefits noted in the previous chapter:

> There are many, many spinoffs from the oil industry. If you have, . . . an oil company in, they have to have all the support amenities that go with it. You have to have fuel, you have to have repairs, you have to have dry docks, you have to have food, you have to have housing. . . . You know, I mean, the spinoffs are tremendous.

Other economic benefits had been enjoyed by the public sector, over a period of decades. While the recent dropoff in oil revenues has shown, in retrospect, the dangers of having had so many of the state's tax dollars in one basket, oil revenues had a long history of permitting other taxes in the state to have been quite low. As Jones (1988, 81) noted about the neighboring state of Texas, the citizens of Louisiana "use oil and gas not only to propel cars and heat homes, but also to educate children, build roads, and provide a variety of public services."

One consistent theme in the Louisiana interviews, however, was that the bust and the boom were very much interrelated—perhaps even inseparable. As was pointed out in almost all of the Louisiana discussions about economic realities, the region's problems in attracting economic diversification, despite its important natural advantages, were related in part to the kinds of changes that had been brought about by the boom.

As one official put it, "It's easy to see why we've gotten so attached to oil [employment]. You just can't get any other jobs around here that pay anywhere near that well. When you're trying to attract another industry, though, they always seem to ask whether they'll be able to get workers if they can't afford to pay that much."

A number of residents noted the degree to which the region had become dependent on oil development; several of them also identified more specific problems. One such problem had to do with the ways in which the previous employment opportunities had affected overall educational levels on the Gulf Coast. As the fabrication and support sectors of the economy grew rapidly, demand rose for certain crafts, such as welding. Because the supply was short, wages rose. By the middle-to-late 1970s, a person could drop out of school, get training in a specialized trade (e.g. argon welding) and make $18–20 an hour. A decade later, as employment opportunities declined along with the drop in the price of oil, this same person had house payments and a family, but few marketable skills, and little education. An economic development specialist put it this way:

> The downside to the oil industry, as I see it, from economic development, has been the deterrent in education. [That's] because so many of our people—high school students, drop-outs, those types—were able to get such good jobs, say as a roustabout working in the oil industry, that they didn't finish their education. They've come back. They've either been sitting on the unemployment roll or we've convinced them to go into other types of training—skilled-labor training.

Another economic development specialist saw a related problem as resulting from the nature of the boom-era experiences of local firms. At the time of the interviews, that specialist was in the midst of an effort to help local firms to adapt to the loss of oil activities by "procuring" government contracts, but the effort was proving not to be a simple matter:

> In fact, to give you an idea, if we'd have had our procurement program in place during the late '70s, early '80s, they'd have laughed at us . . . you go to these manufacturing shops or machine shops, fabrication shops, all of these oil-related industries, they'd laugh at you if you'd say, "Would you like to try to contract with the government, provide goods and services to the government?" And they'd laugh at you. Now, it's a different story because business isn't there. They used to say there's too much red tape. As you know, in the oil industry, they need something yesterday. They call you up and say, you load the thing up on the truck and worry about the paperwork tomorrow. You know, they just sent it out. But with the government contract, you've got to get the paperwork in order first and then you get the contract.

One of the key problems, ironically, had to do with the very level of prosperity that oil had represented. As one observer noted, "It was counter-productive, in that many of the companies that wanted to come in at the

time, they couldn't compete with the salaries being paid by the oil industry.''
Another put it this way:

> In the oil industry it's—well, cost is nothing. You need a generator, you call
> this place. . . . It might cost $500 for a generator. They don't have one, but
> they know where they can get one for you. So they say, sure, we got it. They
> go get it and you pay $1,000 for, the oil industry pays $1,000 for it . . .
> [while the supplier will] pay $500 to get it from somebody else. That's
> [how] it was—there was no quality control. That's one of the problems that
> you find with our mission chart. No quality control. They didn't worry
> about that.

And where did this lead? As one resident put it:

> I think we have a very spoiled business community because things were so
> good. The saying was you could just put a net up across the street in front
> of your building, and at the end of the business day you pulled it in, threw
> all the money in a safe, and went home. The next day, the first one in put the
> net back out. And now these people who are salesmen are really having to
> go out and knock on people's doors, sit down with them and listen to their
> bad jokes. And it's been quite a readjustment for the whole community.

Although there were wide variations in the suggested approaches for
dealing with the region's ongoing economic woes, there was virtual unanimity in seeing the dependence on the oil industry as a problem. The problem
was seen by many as having been exacerbated by the fact that, with the
advantage of hindsight, it had become obvious that many individuals had
not been wise in handling the boom days of OCS development; stories of
economic success, followed by later bankruptcies and other problems, were
in abundance.

Concerns about Ways of Life

About the time when the boom days were ending, the *Wall Street Journal*
(1984) carried an article concluding that extraction had ''depleted'' the region's Cajun culture, as well as the oil, but this would not have been the
majority view among the people we interviewed. If anything, offshore oil development was given greater credit for helping to hold the culture together
than for tending to pull it apart. In important if little-appreciated ways, it appears, the more prosperous days of OCS development may indeed have
helped to sustain locally valued ways of life.

Southern Louisiana's largely Cajun populace, long accustomed to the extraction of the region's natural resources, responded eagerly to the expanded
opportunities for employment and business development. The attractions of
offshore employment went beyond the wages. In particular, there was a good

deal of appeal in the form of "concentrated" work scheduling that evolved in Louisiana, where a person works and lives offshore for a period of time (commonly one or two weeks of consecutive, twelve-hour days) and then is off–duty for a similar period of time. As we noted in the previous chapter, extended versions of this scheduling eventually allowed some Louisiana workers to take advantage of jobs that were thousands of miles away. For much of the offshore oil development in Alaska, for example, a common form of scheduling involved thirty days on and thirty days off, meaning that even the workers on the North Slope oil developments and the Alaska pipeline actually lived in Louisiana, making just one "commute" to work every two months. For most Louisiana workers, however, the key fact was that even the more common 7/7 or 14/14 schedules could allow them to enjoy highly paid offshore employment along the Gulf Coast itself, while continuing to live in hometown locations. Sometimes those hometowns were hundreds of miles from the coast, and often these communities offered few if any other options for employment.

A local newspaper editor, a man having considerable familiarity with the Cajun culture, was one of several persons who noted that this form of work scheduling may have been a particularly important factor in allowing offshore employment to contribute to cultural continuity:

> The principal [benefit] that occurs to me is that . . . the offshore drilling industry is a very vivid kind of industrialization in one sense . . . but on the other hand it is a very benign form of industrialization, because, almost unique amongst kinds of industrialization, it does not require massive disruption and migration in the area in which the industrialization occurs. . . . I think that is the main reason why it has not only been highly accepted here, but has had a chance to interweave itself into the local fabric, because it does not interrupt the local fabric to the very great extent that industrialization that is sort of factory-based would do to a society. . . . People didn't have to all move to Baton Rouge [to get the jobs on the offshore oil rigs] . . . they are still in their same communities.

Concerns about the Biophysical Environment

When it comes to views about the physical environment, it is useful to discuss two separate sets of issues. The first set has to do with the *coastal* environment, specifically including the coastal marshes; the second set has to do with the *marine* or offshore environment. The Louisiana interviews revealed considerable concern about the effects of offshore development on the coastal environment, but very little concern about the effects further offshore.

Coastal Environment. One of the reasons for the concern about the coastal wetlands is that they are disappearing—doing so, according to rough but sci-

Land loss from marsh-buggy use. Note that, rather than regenerating, marsh along the former marsh-buggy track is turning into open water.

entific estimates, at the rate of about fifty square miles every year (Turner and Cahoon 1988). In essence, the entire southern portion of the state of Louisiana consists of soil that has been shipped downstream by the Mississippi river. The Mississippi is called "muddy" because it picks up eroded soil from Iowa and Missouri, and most of the rest of the middle of the country, and carries it downstream. Over the centuries, as the river has periodically flooded and changed its course, it has deposited countless tons of this topsoil across Louisiana's coastal marshlands, usually doing so at the rate of a fraction of an inch per year.

After thousands of years of such "deposition," the net result has been the buildup of thousands of square miles of coastal marsh. So productive are the marshes, in terms of fish and wildlife, that Louisiana has long produced roughly a quarter of *all* the seafood in the United States. As another measure, roughly 90% of the fish species caught commercially in the Gulf of Mexico spend at least part of their life in coastal wetlands and waters.

After decades of apparent stability, however, these coastal marshes have recently started to disappear at an alarming rate. The underlying causes are not yet fully understood, although they clearly include far more than just the offshore development of oil and gas; levees have been built along the Mis-

Land loss along canals. Note that the only land remaining along several former canals (straight lines, upper left of photo) are what were once the "spoils piles" along the canal edges.

sissippi, for example, in the effort to limit flooding, but in the process, they have also limited the extent to which floods could nourish the marshes with new sediment. Instead of spreading across many miles of marshland, the Midwestern topsoil has been channeled out to sea, where it is simply dumped over the edge of the continental shelf. Without new soil, the many tons of wet goo that make up the marshes are slowly sinking, partly under their own weight; any increases in sea level, of course, only make the problem worse.

While OCS development is not the entire problem, a central concern that emerged from many of the interviews was that offshore development in general, and OCS activities in particular, have contributed to the environmental impacts on the state's wetlands. Each of the following paragraphs comes from a different Louisiana resident:

[The offshore oil companies] are not the whole culprit. I blame a lot of the problems in coastal erosion . . . on the regulators. They failed to recognize the impacts of the canals[7]—mainly in the way they were designed and built. They could have built them in such a way that they would have minimized the impacts. In Florida, 30 years ago, [researchers] worked on experiments and studies. . . . Instead of box-cuts [i.e., essentially vertical side-walls for the canals], which completely destroyed water quality, completely destroyed fisheries habitat . . . they used sloping cuts, and [for canal routes, they used] meanders instead of a straight line. . . . Instead of a 100-foot by 10-foot box-cut or 12-foot box-cut, you know, they made the canals 120 feet, tapered at 30-degree angle. . . . The tapered cut allowed them to maintain the riparian zone where fish reproduction takes place, where $\frac{1}{10}$ of the area of the stream produces 90% of the fish—the 10% on either edge where frogs and snakes and all the near-shore critters live, that's what feeds the whole system. The fact that you have a tapered cut would allow your vegetation to grow on the edge . . . [and] instead of . . . increasing energy like a blank wall does, a tapered cut lifts the waves; the energy of the wave is dissipated by the time it hits the bank. . . . It's not the hundred acres you take to build the canal that's so devastating, it's the 10,000 acres you destroy because you build the canal, and then you block the bayous and streams, and you block the tidal exchange.

Well, there are canals and such, but a lot of the onshore development problems that developed was with building canals to reach exploration sites—allowing salt-water intrusion into the fresh water marsh. That tends to kill off the plants that live in the marsh, and when you lose the plants, you lose the marsh.

[If oil development were starting today,] I think there would be much greater attention to the effect on the estuary, even from people whose educational credentials would not lead you to assume they would express things in precisely that way. It would be a lot of concern. . . . I think the gravest change in the way that Louisianians perceive the environment has been in that way—that there is a sense of cause and effect, that it's not just an abstraction that it's bad to pump something into the air because of global warming, even, which is real to some and an abstraction to others—but this is a very real thing, you know. You get these canals and it washes away the land and the shrimp I used to be able to come out here and catch.

Marine Environment. While the *onshore* impacts were seen as having been more significant than was initially recognized, however, relatively few *offshore* risks were seen as serious, save for the personal injuries sometimes associated with the work. A number of concerns were expressed about the ways in which onshore support for OCS activities were managed, but the offshore risk-management record of the industry was perceived as being quite good, particularly with respect to the potential for major oil spills. When asked about the risks of OCS development, one person responded this way:

We haven't had an oil spill—we've had a lot of people get killed. [Pause] I think as far as worrying about an oil spill, I don't know of anybody who worries about that, in the sense that it would be real awful if it happened, but then, we kinda knew that.

Three others offered similar assessments:

As in any industry that develops, you go through the process of hazards on the work place, improvements being made. Louisiana lived through the process of improving the work environment and reducing the hazards. . . . There were a series of occupations that were hazardous, offshore; fishing offshore is not a very safe activity. So you had a lot of people who were used to working in the Gulf and who were used to hazardous duty, and when they got the opportunity to work in the oil and gas industry, this was just another hazardous place to work.

You know, we haven't been as astute in the environmental area. I think we're becoming much more aware of it, and we're working at it much more seriously. I think that such people as the PIRO organization—I'm sure you're all familiar with that[8]—the oil companies have gotten together to come in with strategic centers in case of oil spills and this type of thing.

Now, perceptions of risk, you still have to consider that offshore oil and gas exploration is pretty risky, there's no question about that. But, the risk—you know, the concept of risk is a relative thing. . . . I've been thinking about risk, and how you measure risk, and where have all the big major spills and the big blowouts occurred. Well, they've occurred in what I consider to be highly risky, frontier areas, frontier in terms of exploration, where, in many cases, individuals . . . people were unable to grab hold of the technology and use it appropriately. I think the best example is what the Mexican government did with the IXTOC well blowout. Those people didn't know what they would do. Get everything I can get out of it. Sure there were U.S. contractors there, but you don't do anything without the Mexican government having primary say-so, from what I can gather, in oil and gas exploration. And my understanding is that it was just an absolute disaster technologically. And you look at the Gulf of Mexico, you've never seen anything like that happen in the Gulf of Mexico. I don't think you ever will. It's a different mind-set about a natural resource and . . . [about] how do we appropriately use technology to minimize risk. It's sort of interesting, that last spill in the Gulf was a result of a failure in a pipeline, am I right? A valve gave out or, I don't recall, I think I heard that it was a valve that went out in a pipeline. It was a function of age . . . which is going to happen—but it wasn't a function of misusing the technology or not using it to its maximum capabilities. So the more years of success you have then your perception of risk obviously diminishes.

NORTHERN CALIFORNIA FINDINGS

The situation in northern California could scarcely be more different from that of southern Louisiana. A person moving to the area, most likely by driving some four to eight hours north from San Francisco, would see no evidence of oil rigs, pipeline facilities, or helicopter companies; indeed, with the exception of the city of Eureka and an occasional large lumber facility— including one large and often smoky operation along the coast in Fort Bragg, just a few blocks away from the Eagles Hall where the hearing was held—

there is little evidence of heavy industrialization of any kind. There is also much less evidence of the prosperity that characterizes the coastal regions around San Francisco, as well as the "wine country" of the Napa and Sonoma Valleys just to the north and east of the Bay area, through which at least some travelers might pass as part of a tourist's route toward the northern California coast. Still, there are obvious consolations; most travelers, in fact, tend to be struck by the region's natural beauty, whether they are passing through the deep redwood forests or driving along stretches of California coastline that range from the beautiful to the magnificent.

Given that there has never been any OCS oil production in this region, the kinds of exploration, production, and developmental impacts that characterize the Gulf of Mexico are also absent here. What is clearly not absent, however, is a considerable level of opposition to offshore oil and gas development. Anti-oil materials—often featuring a red circle with a diagonal slash through a black oil derrick—are posted in windows, storefronts, and car windows. Motels and bed-and-breakfast establishments provide explanations of "why we oppose offshore oil development." General stores have anti-oil petitions on their bulletin boards, next to notices of "Kittens for Sale" and "Needed: Ride to San Francisco." Bumper stickers make clear that the owners of cars and pickups share the same sentiments.

So different are the two regions that at one point, one of us noted to the other that "about the only similarity is that they both like Tabasco." Still, there proved to be an important if indirect similarity in the interviews; we encountered such apparent unanimity, although in this case it involved unanimous *opposition*, that we started asking people if they knew of anyone who would be willing to say, even in a confidential interview, that they *supported* the idea of offshore oil development. We found very few; what we did find will be summarized in terms of the same categories used to describe the results of the Louisiana interviews—the economy, ways of life, and the environment. We will follow with a fourth category dealing with an issue that tended not to come up in Louisiana, namely the credibility problem of the offshore oil industry.

Concerns about Economic Prospects

Desires for Economic Growth. We were able to locate just one pocket of support for OCS development, albeit one that was highly localized, centered in the business community and in at least the historic political leadership of Eureka, California. Within this group, as might be expected, the attractions were largely economic, and the views were quite similar to what Molotch (1976) would describe for a typical "growth machine"—the business and often political leaders who see growth as highly advantageous and who thus try to attract it to their communities. As one of these leaders put it, "This city has been a hold-out . . . we're the ones who supported it against all the others. . . . We've got a good harbor here, but a lot of places have good harbors. Nobody's going to ship something in here to have it arrive later at the big markets. So you look for economic development opportunities and options. . . . That makes me a supporter of offshore oil.''

Even in this group, however, support for offshore development seemed to be weakening. As this leader went on to note,

> I'm still an advocate of offshore oil. But its time has come and gone. And you don't fight losing battles. Sooner or later, the need will arise. . . . [But] we had our opportunity; for whatever reason, we lost it. It could have been a good economic boom. If we put it to a vote of the people today, I think they'd vote it down. . . . We've been vocally in support of oil development, but we've seen the handwriting on the wall, and we've let our permits go . . . we were willing to fight it as long as there was a chance, but it's over and done . . . a lot of the stuff on offshore oil, I've chucked it all. [Interviewer: Does that tell us something?] Yeah, it does. You've gotta make room for current issues. . . . Oil is kind of passé.

As another one of Eureka's long-time political leaders noted, the region's newer arrivals tend to be less oriented toward traditional forms of economic growth-promotion than toward environmental preservation. ''Now,'' as he added, ''there are more new people than there are us old-timers,'' a fact with important political consequences:

> Eureka is one of a few coastal towns, if not the only coastal town, that's supported offshore development . . . but I'd have to honestly say our position has become less popular in the last few years. . . . I've run for re-election twice—it was never an issue for me—but last June, it definitely was an issue.[9] My decision has been to cool it . . . local public opinion is becoming stronger against it. . . . There are still many businessmen who support it, but I'd have to think that if it went on the ballot, most people would vote against it . . . Politicians have learned—if you want to defeat anything, tie it to offshore oil drilling.

Concerns about the Volatilities. While a number of the north-coast residents made it clear that they had consciously decided that the environmental and social amenities of the region were more important to them than the potential for greater economic prosperity—a point that will come up again below—the interviews also made it clear that ''prosperity'' was not necessarily the first word that came to mind when many of them were thinking about the potential economic implications of offshore development. Another set of concerns, and one that appeared to be particularly intense, had to do with the Californians' awareness of the kinds of fluctuations in economic fortunes that have afflicted the oil-dependent regions of the Gulf coast.

As one business leader put it, "It would maybe be a boomtown for a while, but it would be a short-term, artificial thing that you would probably never recover from." A business spokesperson in Fort Bragg said, "Oil development has such a boom-bust dynamic. That doesn't help you—it looks like a space needle graph. . . . We don't see that as [being] desirable at all." A generally pro-growth county commissioner referred to his sense that

> oil will come for a short time—and perhaps give a few jobs—but at the cost of other kinds of growth. . . . Tourism here is strong and growing, but people don't come here to lie on a warm, sunny beach. . . . The weather is cold and foggy, so it really is the relatively untouched natural environment that brings tourists here. Proponents like to say, "Look at Santa Barbara"—but it's an hour away from how many million people? We find here that *any* kind of negative publicity [snaps fingers] is enough to keep people away. . . . One story about [our coastline being] "covered with oil" . . . would be death to tourism in this county.

Lack of Blue-Collar Support. It is often the case that strongest support for development will be found not just among growth-oriented business leaders, but also among blue-collar workers who are interested in high-wage employment. At least among the individuals who could be located through the snowball-sample procedures used for this study, however, the level of working-class support for oil development was remarkably low, particularly in contrast to the distribution of views in Louisiana. One executive in the forest-products industry did say he felt that oil development might be seen as a desirable source of employment opportunities among some of the workers he knew, but with that one possible exception, literally none of the persons interviewed for this study who described themselves as representing blue-collar workers, or who were blue-collar workers themselves, expressed any support for offshore oil development.

Why would the level of support be so low? For the most part, the blue-collar workers with whom we spoke, while sometimes having a decidedly extractive orientation toward the forest products of the region, tended to express some of the same sense of personal attachment to the coast as did the other persons who were interviewed; their views, accordingly, may reflect many of the same influences as do those of their neighbors. In addition, given that snowball-sampling procedures tend to lead to the identification of spokespersons and those who are seen as "leaders," rather than those who are seen as "typical workers," it is possible that a statistically representative random sample would document a greater diversity in working-class views than is reflected in the present snowball sample. Research to date, however, has provided very little evidence of blue-collar support for potential OCS development.

In the interviews, the California residents consistently identified three important industries in the region—fishing, logging, and tourism. The persons in the tourism industry were particularly adamant in their opposition to offshore oil, and those involved in fishing were only slightly less forceful in expressing their views. If there would be any significant source of working-class support, accordingly, it would need to come from logging and the forest-products industry. As one informant put it, "At least some of those guys are gonna go along with the logic that, as long as you want to run your chain saw and drive your truck, you'd better have a source of fuel."

We found so much less support for this point of view than we had expected, however, that we asked a number of knowledgeable locals if we were simply failing to locate an important segment of the population, or if instead there might be some other way of understanding why we were finding so few statements of support. One of the Californians explained it in terms of differing social networks:

> We've never really had that sort of undercurrent that you see in Louisiana, where there are a lot of shrimpers who have relatives in the oil business. . . . [What you see here more are] family relationships between people in the resource business, kind of cutting across the conflict of interest. . . . Enough fishermen have relatives in the timber business that they'd explain the issues they're worried about . . . and that sort of overrides the built-in concern of where our diesel is coming from.

One fisherman saw the connections between the fishermen and the loggers as even more direct:

> I know a lot of fishermen who work in the woods. When the fishing gets bad, they'll work in the summer in the woods. . . . They have their own logging companies, even, you know. And they've been in it for a long period of time. . . . Just [in terms of what I understand from] word of mouth, there's a lot of overlap in the area. . . . There are just enough people that mix between the two, and enough contact . . . people who are the more working-class kind of people might be supportive of something if they thought—if there might be jobs coming about—but I think that's kind of tempered by the reality of what they've heard from fishermen who feel they'd be affected.

Concerns about Local Ways of Life

When social scientists refer to "ways of life," they mean far more than "lifestyle," at least if "lifestyle" brings to mind the latest consumer gadgets featured in the supplements to the Sunday newspaper. Instead, the meanings tend to involve a combination of social and cultural factors. Important *social* factors include relationships and ways of interacting with other human beings—

with neighbors, friends, and even relative strangers—involving a relative freedom from fear and crime, for example, or as one Californian put it, "the ability to live in a small, friendly community where just about everybody knows everybody else." Important *cultural* factors literally include the "ways" in which people live and think about themselves, as in a continued respect for Cajun traditions, in many communities of southern Louisiana, or a commitment to live in (and to help "preserve") a relatively unspoiled environment, in many of the communities of the northern California coast.

One of the reasons that "oil is kind of passé" along the north coast is that the economic attractions of oil development, while quite salient in Louisiana, seem not to be that appealing in California—and one of the reasons for the lack of appeal is that oil development might endanger some of the aspects of the local ways of life that are most highly prized. In the view of one Chamber of Commerce executive, for example, OCS activities present at least the potential for "the destruction of our present economy and way of life"—an outcome that "could be brought about even without any oil spills"—while at the same time, "we don't see that the employment promises anything to local workers."

It is partly because of the concern about threats to valued ways of life, in fact, that a growing number of residents of the northern California coast see oil-related growth as being somewhere between undesirable and unacceptable. As one businessman said bluntly, "The vast majority of the people I know in business here came up here to get *away* from that kind of shit." Another business leader noted that, as someone who always had to worry about profitability, she dealt with economic concerns on a daily basis, but she expressed considerable annoyance that representatives from MMS and the oil industry showed so little willingness to discuss the *non*-economic implications of development: "We keep talking about a broader set of concerns, and they keep talking about money. It just shows that they're not listening." Another noted that promises of economic growth would prove to have little positive effect in the region, for the simple reason that "It's the wrong carrot."

To some extent, in fact, northern California proved to be the opposite of southern Louisiana in yet another way, in that additional economic development, in general, was often considered undesirable. The strongest objections, however, were reserved for the cyclical, boom-and-bust disruptions often associated with extractive economies. As will be noted in the next section, Californians expressed concerns about the physical changes that OCS development might bring to the beauty of the region, even in the absence of any kind of oil spill, but they also worried openly about the social changes that the same developments might bring to what they considered to be peaceful, friendly, and largely rural communities. The salience of potential threats to both the physical and the social characteristics of the region tended to be in-

creased by the negative point of reference provided by the heavily urbanized and polluted areas of California, such as Los Angeles, particularly
given that a number of northern California residents had made a deliberate
choice to move to the region to "'escape'' Los Angeles and other such urbanized settings.

Even among the types of business leaders who would normally be pro-
growth, Eureka was the only city in which we found support for offshore
drilling. In Mendocino, a Republican real estate agent said, ''I haven't heard
anything that would make me feel I could support drilling off the coast of
California. . . . We're all so different around here, but we all agree on this
[OCS development] issue.'' A self-described conservative in Fort Bragg effectively agreed: Opposition to OCS development ''is the one issue out here
that cuts across all the lines. And we've got a lot of lines!''

Comments from other business leaders also tended to confirm this assessment. A businesswoman who described herself as an active Democrat
said, ''You wouldn't dream of letting your toilet run into the living room;
putting offshore oil [development] here would be just as out of place.'' At the
other end of the political spectrum, a businessman who described himself as
a member of the Republican Central Committee put it this way: ''I'm generally in favor of almost anything that will contribute to the growth of the
American economy, but that's not absolute. There are some places where certain kinds of development just make no sense, even to people like me. Around
here, offshore oil development just makes no sense.''

In one interview, a city administrator who had complained that the city
and county were ''broke,'' but who had nevertheless expressed little enthusiasm for oil development, was asked a pointed follow-up question:

> [Interviewer:] But you just said the whole county is broke, and here's
> a chance for the . . . administrator to have a source of money that doesn't
> require taxing the folks who vote [here]. . . . You don't sound very inter-
> ested anyway.
>
> [Response:] I'm not. And I don't suppose the county would be either.
> My sense is that they would go broke rather than do that.
>
> [Interviewer:] How do I explain that to somebody in the Minerals Man-
> agement Service? If I go back and simply quote that. . . . They'll say,
> ''Come on, he didn't really say that. You talked to some long-haired liberal,
> right?''
>
> [Response:] People are living here for a reason. I live here because I
> like the area, I like the environment of the area. Yeah, I can be making
> $80,000 or $125,000 a year in L.A. rather than the $50s I'm making here.
> And those opportunities have come along, not just for me as a professional
> but for a whole wide range of other people who live here. We live here be-
> cause this is the place we like to be, and we're willing to take the economic
> risks and economic damage that goes along with it. We're not about to buy

into what we see as damage to the living environment that we have, just to sustain our salary or boost it up. If that's what you want, move back to L.A. to do it.

Not even from a former oil executive could we elicit a statement of support. One of the Californians we interviewed was also someone intimately familiar with the oil industry, having established a seismograph company and operated a number of other oil-related businesses in Texas. While no longer supporting offshore development along northern California, he was able to provide a personal perspective to the differences in regional viewpoints:

> I guess, growing up, just being exposed to Louisiana and ultimately to Texas, Galveston . . . it just seemed like the pollution and contamination was a way of life. Nobody seemed to really give a damn about it . . . because it was the fundamental economic basis for the existence of everything down there, and people just seemed to tune out the pollution problem. . . .
>
> When I started coming up here eight or ten years ago and got to know [some of the people around here who are] ecologically oriented, my initial reaction was, we're stuck with it, on the Gulf coast, why should you all be an exception? But I've undergone an absolutely total metamorphosis since then. . . . Now I've come to feel very deeply that this is just too pretty a place to get screwed up. . . . I say that somewhat with tongue in cheek, because I'm a registered Republican, and I mostly voted Republican in Texas, although I've become terribly disenchanted on this issue. . . .
>
> It was part of my overall metamorphosis to find out that these [California ecological] concerns are real and genuine. I even did some modest oil investing around Lake Arthur, which I'm sure you know, and various places in Texas. . . . I was just exposed to a different milieu, a different environment. People just didn't care about it, or if they cared about it, I suppose they just tried to sweep it under the rug. But out here, it was really astounding to me in the beginning how people did get involved and did take avid stands on real issues. I just never was exposed to that in the south.

Concerns about the Biophysical Environment

As should be clear already, very much related to the concerns over local ways of life were concerns about the physical environment. Unlike the case in Louisiana, where there was an obvious difference in the levels of concern about the *coastal* versus the *marine* environment, the clear tendency in California was to express an almost fervent concern about both, with few residents seeing the two as anything but interrelated.

Many of the interviews, accordingly, wound up focusing on the beauty and the relatively pristine nature of the northern California coastline. While it is clear that cultural, social, and individual factors enter into the definition of what is and is not aesthetically pleasing, as well as into the definition of what is at risk, the physical characteristics of California's northern coastline

California coastline, as viewed from California Highway 1, which parallels the coast for most of the entire state.

are clearly among the reasons why it is widely seen as one of the most beautiful coastlines in the country. The coast appears to be particularly important for the many people who have made conscious choices (often deciding to forego substantially higher incomes elsewhere) to live along the northern California coastline. A significant fraction of the interviews indicated that development would threaten the very environmental characteristics that residents valued most deeply—with threats being presented not only by spills, but also by the visual intrusiveness of platforms, support facilities, and vessels, even in the absence of any kind of oil spill.

Although concerns about potential oil spills were not expressed as frequently as we had originally expected, a number of Californians did raise the issue; when they did so, they generally focused not so much on the predicted likelihood of mishaps as on the fact that they found even the potential for a spill to be unacceptable. Their concerns were not so much statistical as fundamental, particularly in the case of people who still remembered their feelings of revulsion at seeing the painful deaths of birds and mammals in the Santa Barbara spill more than twenty years earlier.

The intensity of the Californians' attachment to the coast was readily in evidence: words such as "rape" and "defile" were used frequently, and a

Louisiana coastline, as viewed from a helicopter, complete with oil-related installation.

large fraction of the Californians referred to the coastline with something akin to awe. They made such references in spite of the fact that most of the persons interviewed for this study were the sorts who were normally more comfortable using words that are more factual, and less evocative, in meaning. Many of them referred to their discomfort in sounding so lyrical, although they often followed by noting that their usual language simply wouldn't convey what needed to be said. As one of them put it, reflecting self-consciously on his own words, "I'm a member of the Republican Central Committee, and I'm pretty conservative, but when it comes to the idea of offshore oil development, I start sounding like the craziest left-wing fanatic around here."

So did many of his fellow residents. Even the kinds of persons who would normally be seen as no-nonsense business leaders would begin to use vivid language when they tried to explain the significance of the coastline to interviewers who were not from the region, referring to the jagged bluffs overlooking clear water, the sight and smell of waves breaking on the rocks, the pleasures of sunbathing on their favorite sandy stretches, or the importance of being able to savor a sense of escape from modern urban life, enjoying a secluded stretch of the coast where there are few signs of traditional human activities other than an occasional fishing boat.

In short, perhaps one of the reasons why our interviews found so little support for offshore oil—even among the groups that are traditionally the most vocal proponents of development in other regions, such as business and particularly real-estate development leaders—may be that the business leaders, too, tended to share the region's widespread concerns about the environment and their way of life. One businessman in Humboldt county told us, "I've started the kind of business that I could have started anywhere; I made a very conscious choice to work and live in this kind of natural setting, and I don't want to see anything that would mess it up the way oil development would."

Even the leaders who were not so fiercely opposed to offshore oil themselves were able to work up little in the way of enthusiasm. As one businessman in Arcata put it, "The people who live around here are adamantly opposed to offshore oil development, and those people are my customers. I don't always agree with my customers, but I don't see any real way that oil development would help me or my business, and I don't relish the thought of the kinds of changes that it would bring to the area, so I guess I'd have to say I'm opposed, too."

Overall, while some of the residents emphasized concerns about the biophysical environment, some concentrated on sociocultural concerns about local ways of life, and others emphasized both, perhaps the most common single theme in the California interviews was that, while offshore development might be appropriate for some regions, they considered it to be completely senseless for the northern California coastline. In a realtor's analysis, "The obvious economic base for our area, besides the fact this is a beautiful place to live, comes from tourism, fishing, and lumber, and oil drilling is not complementary to any of those things." A businesswoman in a nearby town expressed a similar conviction in different words: "It's obvious when you look out the window that this is not a place for oil development."

Another resident put it in broader terms: "What's the difference between Louisiana and California? What's the difference between an apple and an orange? . . . The whole question that you're asking may be a poor way of looking at it. Probably there's nothing in common, there's no comparison. . . . If they're going to have [oil development around here, they] . . . have to approach it on California's terms and not on Louisiana terms." As should be apparent by now, however, California's terms appear not to be highly favorable toward oil development. As one official observed,

> It's not real popular to be pro-oil. I can think of a Supervisors candidate who was pro-oil in the prior campaign; when he ran this time, four years later, he said that he would not support [a proposal to declare the area an] Ocean Sanctuary, but he was no longer pro-oil. [Interviewer: How'd he do?] He

lost! [laughs] People remember! Everybody thought it was going to be close, but he got blown away.

The Credibility Problem

As may already be evident, the California interviews repeatedly revealed a pattern that simply did not arise in Louisiana—a credibility problem that was little short of profound. One of the clearest findings in California was that the claimed advantages of development, and the claims that problems could successfully be avoided, were simply not believed.

Skepticism about Promised Advantages. The credibility problems start with the belief that local workers might not actually enjoy much in the way of significant employment benefits from offshore development. Three main factors appeared to lie beneath this skepticism. The first two, which were expressed mainly by opponents of development, had to do with the specialized nature of job requirements and with what were often described as relatively low estimates for the hydrocarbon potential of the region. The net effect of these two factors, in the view of one resident, was that "There probably won't be that many jobs, they probably won't last that long, and except for a few jobs at the very low end of the pay-scale, they'll probably all go to guys from Texas anyway."

The third factor, by contrast, may actually have been mentioned with the most intensity by the Eureka-based *supporters* of development, a number of whom seemed to feel they had been misled by their intended allies in the oil industry when they went out on a political limb, several years earlier, to attract a fabrication facility. An official in Eureka offered this summary:

> The city of Eureka and the county of Humboldt were working on a project to try to get Exxon to build a platform here. The city owned 120 acres of usable land across the bay, and it's an ideal site because it's on a deep-water channel and there's no obstruction to the ocean—in other words, there's no bridge you have to go under. And because of the size of the platform they were talking about building, we acted as bidder. . . . We felt that we could come in with a competitive bid and, with some political help, be successful. . . . We ended up losing that platform deal—it went to Korea. . . . We have always kind of tried to tie with the oil industry the concept of, okay, do some of your other work here . . . kind of as a trade-off, although it's unspoken. . . . We get a lot of great promises, but . . . I'm not saying the companies are all two-faced, but . . . we still haven't got anything here.

A county official had a harsher assessment:

> The city of Eureka and the Chamber of Commerce were working with Exxon to attract a fabrication yard for an offshore oil platform. There was

a belief that if we were willing to sacrifice our area, that the oil industry would come here and provide some economic benefits; and so the community really went all out. . . . Both the city of Eureka, and the county went all out, too. We did all the environmental work; the city offered land for something like a dollar-a-year lease . . . the county public service department sort of turned itself over to this task . . . We got a sited platform construction yard in California, which many said would be an impossible thing to do; and we did it. And Exxon felt that, apparently, that the cheaper, the lower labor costs were more important than the political support they enjoyed at the time in our community, so they went off to Korea. And I think there was a feeling that we had tried to strike a deal, and that this was a good indication of how much the oil industry could be trusted to do to return some benefits to the community. So I think the doubts about what we had to gain from it grew tremendously.

Then the second thing . . . this company from I think it was Seattle, a shipbuilding company, proposed to open a smaller yard [to build just modules] . . . here in Eureka. And they came and negotiated with the local building trades for reduced labor costs, which they [the trades] gave them, and the city offered extremely attractive terms for property. . . . Then the company went back to Seattle and said, ". . . this is what the workers in Eureka offered. What are you going to offer us in way of concessions?"

The unions up there offered even greater concessions, and the company wound up locating the project in Seattle or wherever it was. And, you know, the feeling of the working people in the community was that we were used. . . . So again, concerns about what was in it for us grew.

Skepticism about Federal and Industry Assurances. As these most recent comments show, even some of the one-time supporters of oil development indicated that they had become more skeptical about the economic promise of OCS development. In addition, however, if any one reaction could be said to characterize the vast majority of the interviews, it would be the conviction that the non-economic assurances from MMS and the oil industry were also not to be trusted. Understandably, given the past history of relationships between MMS and the local area, there was even a good deal of suspicion toward the very idea that the MMS would be supporting our own study of "risk perceptions." One person, for example, simply asked point-blank, "Is this just another PR gimmick?" Such suspicions caused relatively few problems for our own study, perhaps in part because we had come to California to understand local viewpoints rather than to challenge or to ignore them, but it quickly became clear that the skepticism would create significant problems for any efforts to drill for offshore oil.

The credibility problem is one that appears to range from the very general to the very specific. Perhaps the broadest view came from a resident who illustrated her concern in terms of a mosquito spray that had been used in her neighborhood when she was young, but that has since been declared unsafe. "I remember my mother saying, 'They wouldn't do this if it wasn't all right,' and now it turns out, that wasn't the *only* thing that wasn't all right. People are increasingly skeptical about how their government is protecting them, or not protecting them, or lying to them. . . . The Savings and Loan scandal is another example. . . . It's all connected . . . [to the increased skepticism, especially] when they do these stupid things, and these bad things, against our will."

Other concerns were somewhat less broad, centering mainly on the credibility of assurances that oil development could be carried out in an environmentally acceptable way. Many of these comments referred to the phenomenon that, in the social science literature, is sometimes described with the more precise term of "recreancy," or the failure of institutional actors to carry out their duties with the degree of vigor that would be desired (for a recent summary, see Freudenburg 1993). The comments themselves, however, tended to be far more blunt.

The oil industry came in for particularly sharp criticism; in the view of one north-coast resident, for example, "The [oil] industry doesn't have a very good track record, and [since the claims are] coming from the industry, no one believes it—just by virtue of where it's coming from." In another example, a local official offered this example of what he considered to be disingenuous assurances: "One of the oil industry spokesmen was on the radio locally. . . . [He said,] '*Nothing* goes into the ocean, and when it does, we clean it up.' That's a classic example of why nobody trusts them. And people *don't* believe them." Another said, "All we've ever seen has been lip service." Still another said simply, "They've lied to us before, and we know they'll lie to us again."

In the eyes of many, however, the credibility of MMS was not much higher. As one of them put it, "The federal government has shown, repeatedly, that they don't really care about protecting the environment. Every few years, they decide to try the 'nice guy' approach with us, but when it comes down to any kind of regulation that's going to cost the oil companies money, they go right back to their old ways." While the comment had a ring of hyperbole, the person who offered it was able to provide specific examples when pressed. Many others expressed the conviction that, as one observer put it, "They'll tell us anything just to get a foot in the door, but when push comes to shove, they'll let the oil companies do just about anything they want." Perhaps the most common observation was the analogy that, in the words of more than one, "Putting the MMS in charge of protecting the coast is like asking the fox to guard the chicken coop."

Even the most charitable of these broad assessments, while not necessarily questioning the sincerity of the MMS's desires to protect the environment, did express a deep skepticism about the feasibility of the task. In one person's words,

> People do have a good reason to be afraid. They see space shuttles blowing up, they see tankers running aground; technology is something that people are very scared of, and reasonably so. When you put big, dangerous technologies in people's hands, you can't escape the element of . . . human error. . . . It's just an issue that people don't deal with in risk analysis very well. . . . Who hasn't spaced out setting their alarm clock in the morning, or missed their exit on the freeway? Who's so perfect that they haven't done that? That's what happened to the *Valdez*. . . . The third mate was supposed to turn before he got to this one section, and all of a sudden, oops, that light was supposed to be on the other side of the ship, wasn't it? Crunch! At that point they were going hard starboard and hitting [Bligh Reef].

At the narrower or more specific extreme, observations about the lack of credibility tended to reflect not only a good deal of information about the agency, but an effort to differentiate between the individuals they knew, many of whom they found personable and likeable, and the actions of the agency as a whole. Many of the Californians took pains to express generally favorable impressions of a number of the specific MMS personnel working in the Pacific OCS or "POCS" office of the Minerals Management Service. Even the residents having less affection for POCS representatives tended to be somewhat charitable about them as individuals. One critic, for example, noted that she had "nothing against" the Pacific OCS representatives; she simply did not see their assurances as credible, given that they were not sufficiently high in the agency to be able to make meaningful commitments about future actions.

Views toward the top leaders of MMS, however, were another matter. One environmental leader, for example, aimed his criticisms at the persons holding the very highest positions of MMS. "I think MMS people at the POCS office, there's nothing wrong with the people down there. They're doing a job, and we've met with them, and . . . I trust a lot of them. They give me a straight line. It's the leadership. They're still living with the legacy of Jim Watt." Near the end of the interview, when he was asked about the strategy he felt would be most effective for MMS, he gave an unambiguous response:

> I think the conservative element in the MMS is . . . more than any other reason, the reason why there's no offshore oil out here. . . . Strangle the conservatives, is the first thing I'd say, or put a bag over their heads, throw them in the closet, because—this is true of any environmental struggle that's going on, this isn't just offshore oil. If I was looking at this from a purely strategic point of view, I'd *pray* to have an opponent like that.

They're so far off base in terms of understanding what's going on [in northern California] that you don't have to . . . waste any time making that case to the people [i.e., that the agency has little interest in local concerns]. . . .
    A lot of the environmentalists . . . have been smart enough to do what MMS has not done, and that's to get into MMS's shoes and say, what are they thinking, what are their goals, what are their objectives, how are we going to maneuver around it, how are we going to take it into account. . . . What's happened is that they [MMS] have the wrong people [who were appointed during the Reagan administration]. . . . They had ideologues; it was a class of ideologues, is what it was. It wasn't a dialogue between reasonable people but a class of ideologues. . . . They burned so many bridges during the Reagan [years]. . . . They burned every bridge they could burn—it's a scorched landscape. . . . Maybe the question they should be asking themselves is, how are we going to get some of these bridges back that we burned during the Reagan administration?

A number of the more specific references to distrust had to do with the experiences growing out of the effort, a few years earlier, when several environmental groups and some of the oil companies tried to come up with a mutually acceptable compromise for oil development in northern California. As one observer noted, ''To this day, there are people who are not talking to each other within the environmental community . . . because some people went along without compromising, while others wanted to talk compromise with the oil companies. . . . That was the darkest hour for the environmental community. That was as close as they got to actually developing the coast, right there.'' The divisions, however, proved to be for naught, because as he viewed the outcome, ''Some of the oil companies attacked it, and Hodel reneged.''

As a rule, the criticisms tended to focus on the substantive outcomes rather than the procedural niceties of MMS decision-making. As one resident put it, ''Sure, they ask for our comments, but there's no evidence that the concerns are ever addressed.'' Representatives of environmental organizations were particularly clear about this point, and particularly suspicious about the underlying reasons. As one of them put it, ''The MMS is just on a schedule. They take comments, and issue an EIS, but when it gets to that point on their schedule, they go ahead and lease . . . no matter how valid or serious our concerns have been. The message that conveys to the public is, regardless of what your concerns are, this thing is on a track. We're going to go ahead and do it.'' Another, while expressing at least some praise for the agency's recent steps toward openness, suggested that far more needed to be done; the steps taken so far, he concluded, amounted to little more than ''sending a gorilla to charm school.'' Even if resultant changes made it more pleasant for him ''to be in the same room,'' he continued, ''the past behavior of the beast'' made him wonder whether the change in tactics reflected a genuine change, or merely something more cosmetic. Another environmentalist was still more direct, noting, "They're becoming better communicators, but what they have to say is bullshit."

QUANTITATIVE FINDINGS

As was noted earlier, the in-person interviews were supplemented, in California, by the use of a different technique, a simple version of a quantitative approach that is usually known as "content analysis." In essence, this technique involves the counting and classifying the comments that were sent in (or made in person) in response to the draft Environmental Impact Statement, or DEIS, that was the subject of the hearing at Fort Bragg in February of 1988.

As will already be clear from our earlier description of the public hearings, the overall pattern of reactions to the proposed leasing was overwhelmingly negative; to stop there, however, is to miss much of the insight that can be provided by analyzing the comments more carefully. Unlike the case with the interviews, which as we noted in the preface, were prevented by Office of Management and Budget (OMB) regulations from employing the kinds of techniques that would have permitted statistically valid quantification and analysis, the public comments on the DEIS are already part of the public record, and they can readily be quantified.

While it should not be assumed that the comments on the DEIS will be a statistically representative cross-section of views, neither should it automatically be assumed that the comments will be greatly at variance with the findings that might be obtained from a statistically valid sample. The only explicit comparison of which we are aware between a statistically representative sample and an assessment of the comments offered at a public hearing on an environmental issue (Gundry and Heberlein 1984) actually found a reasonable level of similarity. In addition, problems of the unknown level of representativeness of comments would be expected to be most serious with respect to assessing the overall distribution of support and opposition, not with the somewhat more detailed analysis of the distribution of concerns and characteristics across the various categories of persons offering comments. In addition to the "overall" figure, which will be reported in table 1, two additional kinds of information can be gleaned from this study's analyses. One has to do with the distribution of views across *different institutional contexts,* and the other has to do with the *degree of congruence between the content of the DEIS and the concerns raised in public testimony.*

Our content analysis focused on chapter 4 of the DEIS, the chapter providing the discussion of "Environmental Consequences" of the proposed leasing; this chapter alone was 500 pages long. Analysis was also done on all of the written testimony (115 comments, 622 pages in length) and on half of the oral testimony (220 of the 440 oral comments, covering 730 pages of the MMS transcriptions). All comments were analyzed for their basic concerns (aspect of the human environment likely to be affected, types of impacts expected, etc.), for the overall evaluations being put forward (positive, negative, or relatively neutral), and for at least certain characteristics of the persons offering the comments, particularly their institutional affiliations.

The Effects of Institutional Contexts. As might be expected, the comments dealt with an extremely diverse set of topics, covering not just the likely impacts of the sale, but also ranging from complaints about bad writing to predictions about the energy future of industrial civilization. Within the sample, however, a total of 335 comments dealt in a reasonably direct fashion with the desirability of the proposed lease sale. As can be seen from the "total" column on the right side of table 1, the proportion of positive comments was actually under one in ten; just 32 of the 335 comments, or 9.55%, were judged as positive, a figure that is slightly lower than the 34 comments (10.15%) that were neither positive nor negative. The overwhelming majority—269 comments, or 80.30%—were negative.

More telling, however, is the somewhat more detailed breakdown provided in table 1 and figure 1. Almost half of *all* of the positive comments—15 of the 32—came from persons representing organizations in the petroleum industry; only two came from persons who described themselves as being members of general public. Of the other fourteen positive comments, five came from persons representing Chambers of Commerce, five from other (mainly development-promoting) associations, and four from persons representing local governments (three of them from the city of Eureka).

Strikingly, while none of the representatives of petroleum industry organizations were critical of the proposed leases, this was the *only* group for which the proportion of critical comments was less than an outright majority. Even for Chambers of Commerce, in fact, 6 of the 11 comments, or roughly 55%, were actually opposed to the lease sale. Not surprisingly, none of the environmental group comments were favorable, although 3 of the 39, or roughly 8%, were neutral, with 92% being negative. The proportion of negative comments from persons who did not fall into any of the above categories and/or who described themselves simply as members of the general public was even higher; 138 of the 141 comments from this group, or 97.9%, were negative, while only 2.1% were positive.

The Pattern of Concerns. Tables 2 and 3 are necessarily more complex, in that they are intended to provide additional information. Both tables are intended to permit comparisons between the amount of "space" or attention

TABLE 1
Overall Pattern of Public and Organizational Comments on Draft Environmental Impact Statement for Lease Sale 91
(Number of Lines of Text)

	Petroleum Industry Organizations	Chambers of Commerce	Other Associations	Government Organizations	Environmental Organizations	General Public and Other	Total
Supportive	15	5	5	4	0	3	32
(percentage)	93.75%	45.45%	9.26%	5.41%	0.00%	2.13%	9.55%
Critical	0	6	41	48	36	138	269
(percentage)	0.00%	54.55%	75.93%	64.86%	92.31%	97.87%	80.30%
No Position	1	0	8	22	3	0	34
(percentage)	6.25%	0.00%	14.81%	29.73%	7.69%	0.00%	10.15%
Totals	16	11	54	74	39	141	335
(percentage)	100%	100%	100%	100%	100%	100%	100%

FIGURE 1
Levels of Support and Opposition

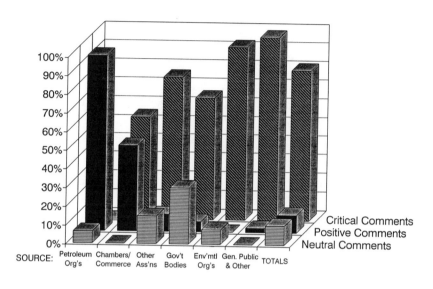

SOURCE: Petroleum Chambers/ Other Gov't Env'mtl Gen. Public TOTALS
Org's Commerce Ass'ns Bodies Org's & Other

Critical Comments
Positive Comments
Neutral Comments

devoted to a series of topics by the DEIS, versus by the various parties offering public comments, as a way of permitting at least a rough assessment of whether or not the issues dealt with in the DEIS have a reasonable level of correspondence with public concerns. The focus of table 2 is on the *aspects of OCS development* that are seen as *leading to* impacts; table 3 deals with *the impacts that are the focus of concern.*

As can be seen from table 2, which summarizes over 30,000 "lines" of comments, the DEIS devotes significantly more of its attention to noise and to oil spills (and to a lesser degree, to congestion and infrastructure) than do the parties commenting on the document. At the same time, the DEIS devotes significantly less of its attention to the impacts of OCS development as a whole. There is a particularly clear contrast against the comments from the general public/other category; the overall issue of OCS development is the source of just 22.9% of the codeable commentary in the DEIS, but 71.7% of the commentary from the general public/other category. The comments from more formal organizations and associations are more specific than those from the general public, tending to fall roughly midway between the DEIS and the general public in the proportions of the commentary focused on the broader or overall impacts of OCS development; this difference appears to reflect the significantly higher proportions of the organizations' comments that were prepared and presented by technically trained representatives.[10] No category

TABLE 2

Topics of Concern: Distribution of Attention in and Comments on the Draft Environmental
Impact Statement (DEIS), Lease Sale 91 (Number of Lines of Comments)

	Noise	Oil Spills	Emissions and Discharges	Congestion and Infrastructure	OCS Development (general)	Total
DEIS Category Totals	1,806	5,190	2,122	3,444	3,739	16,301
	11.1%	31.8%	13.0%	21.2%	22.9%	100.0%
All Other Categories	433	3,328	1,910	1,486	8,618	15,775
(Excluding DEIS)	2.7%	21.1%	12.1%	9.4%	54.6%	100.0%
Petroleum Industry	59	145	62	38	399	703
Organizations	8.4%	20.6%	8.8%	5.4%	56.8%	100.0%
Chambers of	0	21	15	48	99	183
Commerce	0.0%	11.5%	8.2%	26.2%	54.1%	100.0%
Other	58	252	164	234	896	1,604
Associations	3.6%	15.7%	10.2%	14.6%	55.9%	100.0%
Governmental	220	1,510	930	695	2,678	6,033
Units	3.6%	25.0%	15.4%	11.5%	44.4%	100.0%
Environmental	47	702	486	86	1,034	2,355
Organizations	2.0%	29.8%	20.6%	3.7%	43.9%	100.0%
General Public	49	698	253	385	3,512	4,897
and Other	1.0%	14.3%	5.2%	7.9%	71.7%	100.0%

of commentators exceeded or even matched the proportion of space devoted
to noise or to oil spills by the DEIS.

Table 3 summarizes *the impacts that are of concern* to the types of persons indicated, and it also suggests one of the potential reasons for the intense
dissatisfaction that was expressed. As can be seen from table 3 (and the accompanying figure 2, which summarizes the key information in a simplified,
graphic format), there is a reasonable level of correspondence in many ways,
but there are two important exceptions.

First and most clearly, the parties offering comments were significantly
more likely than the DEIS itself to offer comments about the quality of the
human environment. Less than one-quarter of the space in the DEIS was devoted to impacts on the human environment, and perhaps significantly, only
3.8% of the total space was devoted to social and/or cultural impacts—the
"socio-" half of the category of "socioeconomic impacts." The majority of
even this limited discussion, moreover, was devoted to Native American issues, only some of which would have been truly social or cultural if we were
to enforce the definition strictly. The net result, however, is that only 1.8% of

the discussion in the DEIS was devoted to social and/or cultural impacts on affected communities more broadly, specifically including the vast majority of the impact area residents who are not Native Americans.

By contrast, roughly ten times this high a proportion of the discussion by other parties was devoted to this same category of impacts, at 18.2% (16.6% even if comments about Native Americans are excluded), and for the general public at large, the proportion was roughly twenty times as high: four out of every ten comments from the "general public" category—39.9%—were devoted to the "socio-" half of socioeconomic impacts, and even if Native American concerns are excluded, the figure remains at 39.2%. The disproportion for the "-economic" half of the socioeconomic impacts is modest by comparison. The key difference, which can be seen quite readily from figure 2, is that *social* impacts, including impacts on community interactions, local ways of life, and personal well-being, are the focus of more than a third of *all* comments from the general public, but only 1.8% of the attention in the DEIS.

While we have deliberately kept this statistical discussion a simple one, even this straightforward summary should be sufficient to contribute to a pair of conclusions that deserve to be reiterated here. One is that, as was the case in the far more systematic study by Gundry and Heberlein (1984), the overall pattern of attitudes suggested by the public testimony is quite consistent with the findings revealed by the interviews—a pattern, in this case, of overwhelming opposition to offshore oil development in northern California. Not even Chambers of Commerce in the region displayed a pattern of majority support for the proposed leasing, and virtually all of the persons who described themselves simply as members of the public were opposed to offshore oil activities. Second, as indicated by the interviews, the reasons expressed for opposing oil development were ones that went well beyond worries about oil spills, including as well a broad set of concerns about social impacts, community well-being, and regional ways of life that were virtually ignored in the MMS's own analyses. The issue of the reasons behind the Louisiana-California differences, however, is one that deserves more systematic analysis—a task to which we will turn in the next chapter.

TABLE 3
Categories of Impacts: Distribution of DEIS Coverage and Comments, Lease Sale 91 (Number of Lines of Comments)

Source of Topic Statement:	BIOPHYSICAL ENVIRONMENT				HUMAN ENVIRONMENT						
					"Socio-"			"-Economic"			
	Endangered Species	Other Plant & Animal Species	Air & Water Quality	Environment (general) & Other	Native Americans	Way of Life, Community, Well-Being	Land Use/ Infrastructure	Recreation and Tourism	Commercial Fishing	Local Economy, Other	Total
Draft Environmental Impact Statement (DEIS)	1,881 12.2%	6,127 39.8%	1,464 9.5%	817 5.3%	302 2.0%	277 1.8%	557 3.6%	1,319 8.6%	974 6.3%	1,671 10.9%	15,389 100.0%
Category Totals	66.9%				3.8%			29.4%			
General Public and Other	49 1.1%	683 15.0%	304 6.7%	663 14.5%	35 0.8%	1,785 39.2%	72 1.6%	508 11.1%	344 7.5%	115 2.5%	4,558 100.0%
Category Totals	37.3%				39.9%			22.8%			
Organized Entities (Excluding DEIS & Gen. Public)	253 2.6%	2982 31.1%	827 8.6%	938 9.8%	189 2.0%	482 5.0%	621 6.5%	955 10.0%	1,744 18.2%	587 6.1%	9,578 100.0%
Category Totals	52.2%				7.0%			40.8%			
Breakdown of Topics by Categories of Organized Entities											
Petroleum Industry Organizations	14 2.5%	99 17.8%	51 9.1%	144 25.8%	0 0%	16 2.9%	5 0.9%	68 12.2%	138 24.7%	23 4.1%	558 100.0%
Chambers of Commerce	0 0%	0 0%	2 1.4%	23 15.5%	0 0%	17 11.5%	0 0%	44 29.7%	18 12.2%	44 29.7%	148 100.0%
Other Associations	23 1.6%	507 34.3%	46 3.1%	55 3.7%	142 9.6%	43 2.9%	0 0%	100 6.8%	494 33.5%	67 4.5%	1,477 100.0%
Governmental Units	160 3.3%	1,526 31.4%	318 6.6%	485 10.0%	44 0.9%	296 6.1%	564 11.6%	583 12.0%	651 13.4%	227 4.7%	4,854 100.0%
Environmental Organizations	56 2.7%	850 40.8%	410 19.7%	231 11.1%	3 0.1%	110 5.3%	52 2.5%	160 7.7%	99 4.8%	111 5.3%	2,082 100.0%

FIGURE 2
Proportion of Attention Devoted to Three Main Categories of Impacts

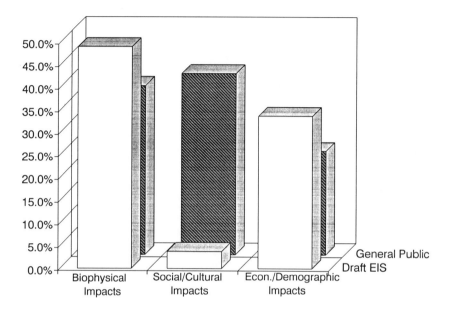

4

Probing the Paradox

Given that there is such a clear pattern of differences in views across regions, how can the differences best be explained? As noted in the Preface, there are a number of potential explanations that are simple and apparently plausible; unfortunately, they are also wrong.

First, unlike many movements that emerge in opposition to developmental activities, the opposition to OCS activity in northern California is clearly not limited to a vocal minority. As should be evident by now, the possibility of OCS development in northern California has mobilized citizen action and cohesion like no issue in the memory of local residents. As one of them put it, "Offshore oil development may be the only thing we all agree on." So pervasive have such changes been that, as another observer noted, California politicians were now tending to run on "motherhood, apple pie, and defending the coast against oil drilling, not necessarily in that order anymore."

Second, calculations about economic self-interest are simply not sufficient to explain the reactions. At least some officials within MMS as well as within the oil industry have argued that Californians oppose oil development because there appear to be so few financial benefits to local communities, while there can be considerable local costs, particularly in providing the types of community facilities and services noted in the earlier discussion of Louisiana's offshore development. In response, these officials have suggested, some form of revenue sharing ought to be offered as a way of counteracting the opposition. Of the twenty-one Californians interviewed in depth, however, only two expressed anything that could be described even as approximating enthusiasm for the idea, and while it may be coincidence, those two happened already to be the two most outspoken supporters of oil development in the entire sample. Among most of the other people interviewed, reactions ranged from lukewarm to hostile. "It probably wouldn't hurt," said one person who was relatively neutral on a number of other points, as well; others disagreed, sometimes strongly, describing the idea as "offensive," "insulting," or "an attempt to bribe us, no matter what you try to call it." Another seemed to summarize the views of a good many of his fellow Californians when he said, "You won't find anybody around here who likes that idea except for our local politicians . . . and they wouldn't actually take the money, because they know if they did, they'd be dead meat.'' All in all, while both of us have had enough experience in working with rural governments to know their need for additional revenue sources, and both of us believe that a greater sharing of revenues with affected jurisdictions would make good policy sense if only on an equity basis, the interviews provide no realistic basis for expecting that such benefit-sharing approaches would lead to a dramatic decrease in opposition.

Third, a number of representatives of MMS and the oil industry have argued that the opposition results from ignorance about the "real" risk of an oil spill—an argument that is political as much as factual, but that deserves a bit of additional clarification even before we examine its applicability. The proponents of offshore development claim, basically accurately, that exploration and drilling activities, in the United States, have a good record of avoiding spills—essentially a clean record ever since the infamous blowout at Santa Barbara more than twenty years ago. It is also true, as critics respond, that the rest of the story is not nearly so rosy; the IXTOC spill was the biggest ever, and it took place in drilling for offshore oil in Mexican waters, not that far from the outer continental shelf of the United States. In addition, the *transportation* of oil has proved to be not nearly so successful, as shown vividly by the *Exxon Valdez*—although far greater quantities of oil are "routinely" spilled at sea every year (National Research Council 1985). The response of the supporters is that if the oil from offshore development were to be brought to shore via pipelines, this would actually cut down the amount of oil that needs to be imported and brought to California in tankers, thus *reducing* the risks of oil transportation. The response of the opponents, in turn, is that oil industry representatives made similar claims about pipelines for serving some of today's oil fields near Santa Barbara, but that at least some oil companies then spent a decade trying to avoid the requirement that, in fact, they build the relatively few miles of pipelines that would be needed to avoid oil tanker traffic to serve the fields.

As is probably clear already, the argument actually involves far more "inside information," selectively applied, than seems on the surface to be the case; more important for present purposes, however, is that it has very little to do with the pattern of attitudes we encountered. Neither "ignorance" nor a predominant focus on oil spills could be said to characterize the opposition of the people we interviewed. Many of the interviews were with people who hold positions of formal and/or informal leadership, and the characteristics of such people should not be assumed to hold for the public at large. Still, these leaders showed an impressive grasp of issues, were well aware of the differences in risks between tankers and drilling platforms, and expressed concerns that went well beyond the risks of oil spills. In this context, it is also worth repeating that, rather than focusing single-mindedly on oil spills, the comments on the Draft Environmental Impact Statement were actually *less* likely to focus on oil spills than was the DEIS itself. Similarly, although the interviews were conducted (and intended to be analyzed) on a qualitative rather than a quantitative basis, a bit of rudimentary quantification is worthy of note: Literally none of the people we interviewed identified oil spills as their sole concern. For the majority, catastrophic spills from oil exploration and development were a source of less concern than were other considerations, including the more chronic impacts on both the human and the biophysical environments from routine oil development operations.

To the extent to which the interviews *did* encounter the issue of oil spills, moreover, the context tended not to be one of describing the probabilities as extremely high, but of describing the credibility of the official probability estimates as extremely low. Several comments, for example, reflected reasonably detailed knowledge of local conditions, such as rough seas or the fact that even experienced fishermen along the northern California coast continue to lose their lives almost every year; others focused on accurate understandings of the shortcomings in the current state of the art in risk assessment. As one person put it, "the official estimate was that a spill the size of the *Exxon Valdez* was so unlikely that it was supposed to occur just once in every 250 years or so, but they only managed to get in about a dozen years before it happened . . . Nobody I know believes those numbers, because everybody knows they don't take account of human error."

Finally, perhaps the most common explanation for opposition to development that involves potential risks—often known as a "locally undesirable land uses," or LULUs (Popper 1981)—is that the opposition is "just" a reflection of a not-in-my-back-yard or NIMBY reaction. We frankly expected to find more support for this explanation than we did. Instead, as should already be clear, the interviews revealed an impressively broad range of concerns; if any single theme emerged, it might have been not so much the narrowness of the people we interviewed, but rather their criticism of *the federal government's* narrowness and short-sightedness. A common response was that OCS development in northern California "makes no sense," when instead of simply exhausting a finite resource at an accelerated pace, the nation would be better-served by an energy policy that would lessen our dependence on nonrenewable fuels.

Rather than expressing narrow, back-yard concerns, in other words, the Californians who were interviewed for this study tended to emphasize broader questions of national energy policy that they saw as being inadequately taken into consideration in federal decision-making to date. "If we're really running out of oil," said one, "then the solution isn't to drill the last places on earth or to do everything you can to use up what's left even faster. It's to use the energy resources we have now to buy time, while we make a serious effort to develop solar power, renewables, and conservation.'' In another, more succinct summary, the current approach ''is like the Surgeon General saying, 'We've got a drug problem—we've got to make sure everybody gets plenty. We've got to get some more of this stuff.' ''

THE UNIQUENESS OF LOUISIANA

In short, there are problems with all of the ''obvious'' explanations of why there would be such striking differences in views toward offshore oil development between California and Louisiana. Particularly if we are to learn lessons that will be relevant for the future, whether for coastal prospects in California or elsewhere, what is clear is that the question needs to be analyzed, rather than simply avoided. What the analysis requires, in turn, are explanations that go beyond those that are ''obvious'' but wrong.

Perhaps the most useful perspective, in this context, is one that comes from observers having a great deal of on-the-ground experience with offshore oil, namely the residents of coastal Louisiana. A number of those residents, when asked for their thoughts about what led to the striking differences between the California and Louisiana reactions, turned the ''obvious'' logic of the oil proponents, and of our interviews, on its head. We orginally expected that the Louisiana residents would follow the common pattern of assuming their own experiences to be ''normal,'' while the experiences of other regions were ''different.'' Instead, the Louisiana residents consistently spoke not about what made the *California* experience so unusual, but about the need to understand that the region with ''unusual'' relationships to the oil industry is *Louisiana*.

They have a good point. It is southern Louisiana, after all, that is home to the greatest concentration of offshore oil activity in the history of the planet. Southern Louisiana is the region where offshore oil first developed, where oil continues to play an important role in virtually all corners of the economy, and where the influences of offshore platforms are felt dozens or even hundreds of miles inland. In Louisiana, even the state bird, the brown pelican, was once driven to extinction in the state, due in large part to the effects of petrochemical products and byproducts, principally DDT. It is the historical uniqueness of Louisiana, accordingly, that deserves more attention than has been evident in oil policy debates to date.

As we have examined the special case of Louisiana, moreover, a significant number of differences have emerged. Louisiana's early experience with offshore oil, in particular, provides a strong contrast against more recent experiences not just in California, but also in many other coastal regions of the United States. While these differences could be discussed at some length, they can also be boiled down to three broad categories of differences—differences in *historical* factors, in *biophysical conditions*, and in *social and economic conditions*. These are the categories that will be used, accordingly, in the following discussion, which draws heavily on Freudenburg and Gramling (1993).

Era of Initial Development

Four key historical factors contributed to the uniqueness of the development of Louisiana's offshore oil and gas industry. The first and perhaps the most obvious of them has to do with the historical era when the offshore developments first began: one of the clearest differences between Louisiana in the 1930s and any other coastal regions in the 1990s is that more than half a century has passed in the interim. Countless changes have taken place in virtually all areas of society, and it would be surprising indeed if those changes were not to have important implications for the reactions that could be expected.

In particular, one of the key changes since the time of the initial offshore development in Louisiana has been a significant growth in environmental awareness across the United States (see, for example, Dunlap 1987, 1992). In contrast with recent years, during which the growth of opposition to OCS activity has become more apparent, the 1930s and 1940s saw the growth of technological "exuberance" (Catton and Dunlap 1980, 15; see also Dunlap and Catton 1979), a time of almost unprecedented faith in technology nationwide. During such an era, there was little conception of marshes as being fragile and finite environments; drilling procedures could be limited to the "most efficient" way to get a job done in the short term, often involving the construction of canals through the marsh to allow submersible drilling barges to operate. The effects may show up in other ways as well: According to Catton's calculations (1989, 110), based on 1983 data from the Conservation Foundation (1987, 20), Louisiana's estimated per-person generation of industrial hazardous waste exceeded that of any other state, at more than three metric tons per person per year, and according to subsequent calculations by the World Resources Institute (1993, 223), the state has continued to rank first in the country, without adjusting for population size, as of the most recent data available from the federal government's Toxics Release Inventory.

Given the relatively low levels of environmental concern that prevailed during the decades when the offshore industry was being established in Louisiana, it was possible for offshore drilling to evolve in the state as an environmentally insensitive activity. It is quite likely, however, that if another part of the country had been the location of the initial push offshore in the 1930s and 1940s, the results would have been comparable. The limited amount of drilling off the coast of Florida in the 1940s, for example, appears not to have differed all that significantly in terms of environmental protection from the drilling that was done in Louisiana during the same era, and the initial drilling along the California coastline at the beginning of the twentieth century was environmentally insensitive as well. By contrast, if oil and gas development were only beginning to take place in Louisiana marshes today, it is quite unlikely that the same degree of freedom from environmental constraints would be in evidence.

Temporal Priority of Oil and Gas Development

Significantly, however, not only did OCS development take place during an earlier era in Louisiana, but it also took place before a number of potentially competing uses had become established. In contrast to proposals for OCS development in regions such as northern California, which are under heavy fishing pressure already, the initial development of offshore oil in Louisiana took place at a time when the state had little prior tradition of offshore fisheries. The primary extractive activities in coastal Louisiana prior to World War II involved the coastal marsh and contiguous swamps (see Comeaux 1972). Shrimp, the most significant of the commercial species in the Louisiana Gulf today, were thought to be present only in the estuaries, and were not even known to be available in the deeper Gulf until the late 1940s (Morgan City Historical Society 1960).

As a result of the temporal ordering of development, both of the major offshore activities in Louisiana—oil and gas development, and the harvesting of fish and other types of renewable resources—essentially grew up together. Early exploration for oil even used shrimping vessels that were leased by the oil companies (*Oil Weekly* Staff 1946). In cases where oil spills have occurred in the Louisiana Gulf since those early days, moreover, they have affected only local areas. Given that much of the fishing/shrimping fleet is highly mobile, working off the coasts of Florida, Alabama, Mississippi, Louisiana, and Mexico, the spills that have taken place have not materially affected harvest levels.

Incremental Onset and Evolution

Third, OCS activity in Louisiana occurred as a gradual extension of land-based activities, first through the coastal marshes and then into ever-deeper federal waters. As noted above, the earliest drilling rigs in the marshes involved little more than the construction of traditional drilling equipment on pilings and barges. It was not until after the passage of the Outer Continental Shelf Lands Act in 1953 that the technology began to evolve more rapidly, with jack-up and semi-submersible drilling rigs replacing the submersibles in

Louisiana oil platforms, old and new. Platform in foreground
is an older one, connected to land by a dock. Note from the
waves that this photo was taken on an unusually stormy day.
Structures in background are all related to offshore oil as well.

the deeper waters. The evolution of offshore technology was paralleled by the
similarly gradual emergence of support services and by altered forms of work
scheduling (seven days on and seven days off, etc.) that were adaptations to
the logistical problems associated with operating at increasingly remote sites
(cf. Gramling 1989). The technology continued to evolve and grow during the 1960s and early 1970s—so much so that even the rapid growth after
the oil embargo of 1973–74 seemed to be a relatively natural extension of the
development that had already taken place. While the northern waters of the
Gulf of Mexico have become the most heavily developed offshore area in
the world, in short, the development has taken place one step at time.
Clearly, such a history is very different from the context that greets proposals
for development elsewhere on the OCS today, where proponents call for the
kinds of massive, technologically sophisticated, and capital-intensive developments that, as will be recalled, the *North Coast News* viewed as representing a potential "invasion" of oil activities.

Local Development and Adaptation of Technology

Fourth and finally, development in Louisiana has always been a local affair.
In general, it has been in the Gulf of Mexico that the offshore technology has
evolved, meaning that developmental efforts in the offshore industry have of-

Construction of Platform Bullwinkle, the largest steel-jacketed offshore platform in the world. Note size of people for comparison. Today this platform stands in over 1,300 feet of water in the Gulf of Mexico.

ten focused on the solution of local problems. Because offshore technology emerged and evolved in the Gulf, moreover, not only is there considerable familiarity with the activity, but there is also a pride in the technological accomplishments that have been locally achieved.

In sum, offshore oil development in Louisiana began when environmental awareness was quite low by contemporary standards, and during an era before a number of potentially competing uses had been instituted, with the industry thus becoming established either before or along with a number of other industries in the region. While today's OCS developments in the Gulf are massive, they did not start out that way; they evolved gradually, and locally, always as what would have appeared to have been relatively natural extensions of prior experience. It stands to reason that a different reception would be expected for proposals that are made during a time of high environmental sensitivity, that pose some threat of conflict with established uses that are already experiencing other pressures, and that appear to have been thrust suddenly and massively into a region where they seem to be not only alien but ill-adapted.

BIOPHYSICAL CHARACTERISTICS

In addition, there are two sets of particularly striking differences in terms of biophysical characteristics between Louisiana and California; in general, the Louisiana conditions differ from those that are found in other coastal regions of the United States, as well. The first set has to do with coastal topography, and the second has to do with marine topography; we will discuss them in that order.

Coastal Topography

As will be clear to anyone who has visited the region, the coastal topography of Louisiana tends to be quite different from what is found in the rest of the country, but for the purposes of offshore oil development, the key differences fall into three categories—the marshes that limit land-based access to the coast, the estuaries that offer many opportunities for the development of harbor space, and the low relief and energy levels that characterize most of the coast.

Coastal Marshes. As even a cursory examination of an atlas will reveal, the distribution of populations and roadways in Louisiana is very different from the distributions that are found in most coastal states in the United States. Virtually none of the population lives on or near the coast, and particularly in the central stretches of the state's coastline that provide the staging area for offshore oil development, it is often effectively impossible to get within *ten miles* of the coast by road. The reason is simple—most of the "coast" is lined with a broad and virtually impenetrable band of coastal marshes, which form what the *Environmental Almanac* calls "the largest continuous wetland system in the lower 48 states" (World Resources Institute 1993, 172). In general, the marshes are far better-suited for the abundant fish and wildlife of the region than for the humans who need to live on land that, at least by local standards, is relatively "high and dry." As one resident of Lafayette explained, "The Gulf is only about fifteen miles south of here, but there are probably more people in this town who've seen the Gulf from *Florida* than who've seen it from anyplace in Louisiana."

In most coastal states of the United States, but particularly in states such as California and Florida, where proposals for OCS development have met with intense opposition, the situation is virtually the opposite: most of the population lives on or near the coast, and with few exceptions (e.g., the stretch between Legget and Petrolia in the northern part of California), virtually all of the coast is readily accessible by road. In California and much of the rest of the coastal Unites States, the coast is seen as a valuable public

California coastal access. Recreational vehicles, parked and ready to enjoy the sunset, along California Highway 1.

resource, a thing of beauty, and a source of popular recreation—and in part, this is because the coast is actually "seen" by so many people, so often. Even in Alaska, while there are relatively few miles of coastal highways, this is partly because there are relatively few highways of any description; a significant fraction of the state's highway system is literally "on" coastal waters, in the form of ferries that ply what is officially known as "the Alaska Marine Highway." At least by Alaska's standards, the coast is thus relatively accessible, and it is viewed both as a resource and as an important recreational feature. By contrast, one Louisiana respondent noted, "We don't have beaches, we have estuaries. You couldn't go walking along the estuary and find a dead bird even if there was one." Because of the coastal marshes in Louisiana, not only is the coast rarely seen, but local residents' descriptions of their state's coastal regions are more likely to involve mosquitoes and alligators than spectacular visual imagery.

By way of quantifying these relatively qualitative observations, figure 3 summarizes the proportion of the population that lives in coastal regions, and the proportion of the coastline that can be reached by road, in the states of Louisiana and California. To give a better indication of which state is the more "unusual," we have also provided comparative data for Florida, an-

Louisiana coastal access. This is literally the end of the road, at the point of closest road access to the Louisiana coast, between Morgan City and Houma, Louisiana. Here we are approximately 20 miles from the Gulf of Mexico, separated from it by the Gulf Intracoastal Waterway and by miles of open marsh.

other state in which the proposals for offshore oil development have proved to be highly controversial. In all three states, detailed road maps have been used to identify coastal regions having roadways within one mile of the coast; population figures have been computed simply by adding the number of people living in counties (or parishes) that are in contact with saltwater. Despite the fact that this definition is relatively generous—including, for example, the population for Jefferson parish, which lives almost entirely in the New Orleans region, many miles from the Gulf—figure 3 shows that less than a quarter of the Louisiana population lives "along" the coast, while comparable figures for California and Florida are 61% and 76%, respectively. Under a more careful measurement, if Jefferson parish is removed from the calculations, the Louisiana coastal population drops down to less than 11% of the state's total population. Similarly, only 12.26% of the Louisiana coast is accessible even by rudimentary roads, while the figures for California and Florida are 90% and 74%, respectively.

Estuaries. A second key feature of Louisiana's coastal topography is the presence of an extensive estuarine system. While it is difficult for humans to

FIGURE 3
Differences in Coastal Access

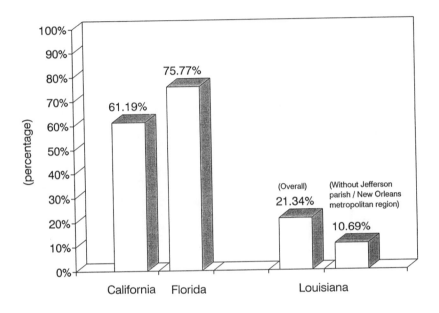

reach the Louisiana coast from land, access from the Gulf is considerably simplified by the numerous bayous of the region; unlike many coastal areas in the United States, Louisiana is characterized by an abundance of waterways that intersect the highway network further inland, and that provide coastal access for marine interests.

This feature of the Louisiana coastline is particularly noteworthy in contrast to the conditions of California: as becomes immediately apparent from a comparison of the two states' coastal charts, access by water in Louisiana is as easy as access by land is difficult, while precisely the reverse is true in California. While limitations on available dock space can be a problem in any coastal region, the physical characteristics of the estuarine environment mean that the most significant limitations in Louisiana are imposed not by lack of suitable harbors, but by lack of facilities—a shortage that can be remedied, and often has been, by relatively straightforward construction projects.

FIGURE 3 (*continued*)

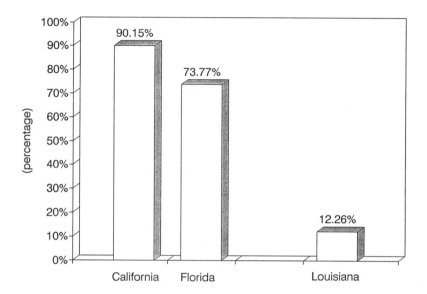

Percentage of Coastline with Roads

Low Relief and Low Energy Levels. A third key feature of Louisiana's coastal topography is that it is a region of extremely low relief and generally low energy. Perceptions of the coast are radically affected by the differences in coastal elevations and the energy levels of the wave action along the beaches.

When Louisiana residents actually do reach the shorelines of their state, they see flat water and flat land, with comparatively low-energy beaches. They see water that is seldom clear because of the discharge of silt by the Mississippi and Atchafalaya rivers. They may also see a major problem with marine litter—including plastic milk and water containers, onions, oranges, bottles of all types, and other garbage, all of which is regularly thrown over the side of offshore support vessels, sport-fishing boats, and other ships in the Gulf. As a simple if concrete illustration, one of us walked slightly less than one mile along the beach at Grand Isle (the only barrier island accessible by road in Louisiana) in November of 1990. Using a rope that was also found on the beach, he "strung" on the rope *only* the gallon-size, plastic milk and

California port facilities. Note lack of remaining space.

water containers along the route; the resulting string of containers was approximately 20 feet long.

In northern California, residents frequently observe the power of the sea where it meets the coast, and they often comment upon it with what they themselves call "awe." In contrast, the Gulf of Mexico off of Louisiana is generally a more sedate body of water during normal periods of time, and even when the Gulf does inspire awe, as during tropical storms and hurricanes, it provides displays that have rarely been seen by people who are alive today. As noted above, only a small percentage of the population lives on the Gulf, and a combination of modern weather forecasting and centuries of experience have led to routine evacuations of coastal and low-lying areas as tropical storms approach.

Marine Topography

Persons who are new to OCS issues often express bewilderment that fishing interests in California, Alaska, and elsewhere, would have such high levels of concern over OCS developments, given that marine use conflicts have been so notably absent in the Gulf. After all, the fisherman who was tied to an oil platform in the Gulf, as noted in the first chapter, is someone who has a good deal of company in the Gulf of Mexico. In New Orleans, the recently completed "Aquarium of the Americas" includes a large display, complete with

Port facilities, Bayou Lafourche, Louisiana. Note remaining space availability; also note use by both commercial fishing vessels and offshore oil support vessels.

a title of "From Rigs to Riches," pointing out the *advantages* of oil platforms as a form of habitat for many of the fish species of the Gulf. The display was made possible by funding from oil companies, but the advantages it summarizes are real ones. Even the two of us have both fished around oil production platforms, doing so "on purpose."

Given the Louisiana experience, why would there be so much opposition to oil structures from fishing interests in other regions? The reasons are not so obscure as they first appear. In addition to the historical factors just noted, there are two key aspects of the marine environment that have helped to limit the potential for conflicts between OCS development and other marine activities, particularly fishing, off the coast of Louisiana. Both of them involve the sea-bottom topography; one has to do with the *gradual slope* of bottoms along the central Gulf of Mexico, and the other has to do with the presence of *silt bottoms* that limit the number of obstacles likely to be encountered.

Gradual Slope. Off the coast of Louisiana, the topography of the continental shelf presents very different conditions from those that are found in areas where OCS development has been most contentious. In most areas of the

FIGURE 4
Differences between California and the Gulf in the Width of the
"Continental Shelf"

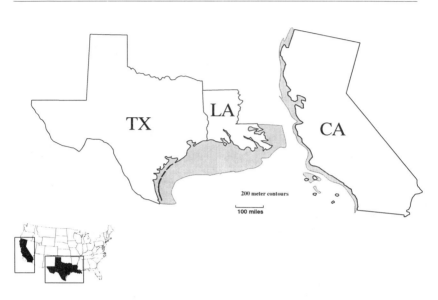

country, but particularly along the Pacific Ocean, the continental shelf drops off much more dramatically into the ocean basin, meaning that it is also much narrower than the shelf that exists in the Gulf. In the Gulf, by contrast, production platforms are in place well over 100 miles offshore, and in some areas, the slope is as gradual as one or two feet per mile. The same Aquarium of the Americas discusses one platform that was built 100 miles offshore (near the Flower Garden Banks, the northernmost coral reef in the Gulf), where it stood in just eighteen feet of water.

Figure 4 illustrates the difference, presenting a side-by-side comparison based on the 200-meter contour lines of the two coasts, with both regions being presented with the same scale of miles. The shaded areas inside the "isobath" or contour line for each region represent areas where the water is less than 200 meters deep. The 200-meter depth is about the amount of water that would be needed to cover a sixty-story office building; more relevant for purposes of fishing is the fact that this is also enough water to assure virtual darkness, and hence a limited presence of plants and other marine life. Waters of this depth can be reached within just a few miles of most of the northern California coast, while being over a hundred miles out to sea for the oil-rich region along the central and southwest Louisiana coastline.

As a result of this difference in sea-bottom slopes, the available area of the Louisiana shelf is far larger than is the case in California, reducing significantly the likely intensity of use conflicts. The gradual slope also reduces the number of problems that are likely to be created by any given obstacle: Even if a fishing boat needs to make a quarter-mile detour around oil operations, for example, there will be little significant impact on the boat's ability to keep its nets in contact with the seafloor. In areas such as the Pacific Ocean off the California coast, by contrast, the steeper slope makes the Louisiana experience largely irrelevant, and in two ways. First, the actual area available for use is smaller, meaning that even an apparently small loss of area for OCS activities can have proportionately major impact on the area available for fishing in a given region. Second, given that bottom-dragging operations need to work along a contour line, following sea bottoms at a given depth, the presence of an oil platform can mean that fishing boats would need to make a detour that would correspond to a difference hundreds of feet in water depth, effectively precluding the option of "fishing around" many such structures.

Silt Bottoms. An additional feature of the marine environment of the Gulf of Mexico is the presence of silt bottoms. While the heavy silt discharges by the Mississippi and Atchafalaya rivers mean that the water is seldom clear, further reducing many concerns over aesthetic impacts, the nature of the bottom also means that fishing operations encounter few obstacles such as rock outcroppings that would lead to the loss of nets and other gear. In regions such as California, by contrast, the frequent presence of rocky outcroppings can severely limit the ability of fishing boats to change their trawl runs.

One consequence of the difference in bottom types is that—to reiterate a point that often comes as a surprise of the *opponents* of oil development in other regions—oil rigs actually *can* provide a significant advantage for fishing operations on silt bottoms of the Gulf of Mexico. Certain types of commercially important fish can only survive in the kinds of habitat known collectively as "hard" substrate—rocky bottoms, reefs, rock outcroppings, and the like. In the central Gulf of Mexico regions of the United States where oil development activities have been the most intense, natural outcroppings of this sort are so rare along the predominantly silty bottoms that oil-related structures have been estimated to make up roughly a quarter of all hard substrate (Gallaway, 1984). In effect, the oil rigs thus serve as artificial reefs, concentrating and probably increasing the fish populations, and it is quite common to see fishing boats literally tied up to oil rigs in search of fish.

In Louisiana, in sum, the coastline is inaccessible to most land-based populations, and accordingly low in social salience, while offering enough potential harbor space to meet the needs of offshore oil development as well as of potentially competing uses such as fishing operations. The offshore sea-floors, meanwhile, tend to have such gradual slopes as to offer vast areas of virtually level bottoms, relatively free of obstacles, but also so devoid of nat-ural reefs that the oil rigs provide a valuable service for fishing operations. As was the case for the historical factors, most of these characteristics are almost precisely reversed for the coastal regions of much of the nation, and partic-ularly for the northern California coast; the very considerations that have contributed to the ready acceptance of offshore oil in Louisiana tend to exert just the opposite effect in California.

SOCIAL FACTORS

Along with the historical and biophysical factors noted above, four sets of social characteristics of the Louisiana population appear to have encouraged the easy acceptance of OCS activities in the 1940s. They involve the average educational levels, the patterns of social contacts, the importance of prior ex-tractive industries, and the potential for overadaptation that characterized the southern regions of the state in that era.

Education Levels

Studies tend to find such broad support for environmental protection in the United States that few sociodemographic predictors show strong correlations with environmental awareness and environmental concern. Contrary to early speculation, for example, recent studies show that blacks are as supportive of strong environmental controls as whites (Mohai 1990), and poor people tend to be as supportive as are wealthier ones (Morrison 1986; Van Liere and Dun-lap 1980, 1981; Mitchell 1979; Freudenburg 1991b; for a more analytical re-view, see Heberlein 1981).

One of the few consistent exceptions has to do with educational lev-els, with better-educated persons in the Unites States generally expressing somewhat higher levels of environmental concern (Van Liere and Dunlap 1980). Thus it may be significant that, particularly in the 1930s and 1940s, coastal Louisiana had some of the lowest educational levels in the country. For example, in St. Mary parish, the scene of initial OCS activity, only 47.2% of the adult population had as much as *five years* of education in 1940, and only 12.2% had graduated from high school (U.S. Department of Com-merce 1940). Other rural areas of southern Louisiana had similarly low ed-ucational levels. By way of comparison, well over 75% of the adults in the United States had a high school education, or more, by the time of the 1990 Census.

Extractive Uses of the Coast

The other industries that most characterized coastal Louisiana at the time of initial OCS development were primarily "extractive" ones; like oil development, that is, they involved the extraction of raw materials from nature. As noted earlier, local residents obtained products both from the Atchafalaya Basin (cypress lumber, fish, crawfish, water fowl, and moss for furniture stuffing) and from the coastal marsh (furs, shrimp, oysters). The export of such raw materials had provided the mainstay of the economy in coastal Louisiana for almost a century prior to OCS development. In addition, as noted earlier, the state of Louisiana was actively and aggressively marketing the offshore waters for oil production prior to the OCSLA.

As a general rule of thumb, persons who are involved in extractive activities—particularly in activities that are consumptive, nonrenewable, or carried on at too fast a pace to be renewable, as tended to be the case for a number of the pre-OCS activities in Louisiana—will be less likely to object to new extractive industries than will persons in manufacturing or service industries. They tend to be far less likely to object than will those whose livelihoods depend directly on the maintenance of high environmental quality. Extractive industries, however, have been shrinking rapidly in their relative economic significance. In percentage terms, the proportion of the nation's labor force involved in extraction has dropped by nearly two-thirds since 1920—roughly as sharp a decline, on average, as the much better-known decline in the proportion of the work force engaged in farming (for further discussion, see Freudenburg 1992a).

Both in northern California and, increasingly, in most coastal regions of the United States, the economy has come to be far more dependent today on the amenity values of the coast than on its extractive values. In addition to the fact that the proportion of the population depending on the extraction of coastal resources is both small and shrinking, the key extractive industry of fishing has moved increasingly toward sustainable levels of harvest, with a corresponding growth of the recognition that the industry is inherently dependent on the quality of habitat. Tourism, the major economic use of the coast in California, is also dependent on the coast remaining relatively pristine. In California's coastal areas, and increasingly, in other areas as well, the likelihood of finding support from extractive workers can thus be expected to continue to decline.

Social Interaction Patterns

A powerful if often unnoticed influence on a person's attitudes comes from what sociologists call social networks and interaction patterns—that is, from the other persons with whom that individual interacts. Even if a given individual does not work in the offshore oil industry, for example, her attitudes may be affected by whether or not her friends and relatives do. Given the historical, biophysical, and other social factors summarized above, the "average" resident of coastal Louisiana in the 1940s would be expected to have known many friends and neighbors who were employed in the oil industry. By the 1980s or 1990s, it was virtually impossible to live in southern Louisiana and not to know *someone* who was so employed.

Over the past several decades, moreover, few *new* persons have moved to the oil regions of southern Louisiana except for those who have been drawn there by the extraction-related prosperity. There is very little likelihood, accordingly, that the in-migrants' attitudes toward oil development would have been any less supportive than would the views of the Cajuns who preceded them. As noted by one observer who was particularly knowledgeable about the region's culture and history, the effects of the social interaction patterns in south-central Louisiana appear to have been compounded, in an almost synergistic way, when combined with the region's historically low educational levels and its distinctive culture:

We have a strong tradition here of patronage, meaning not in the simple political sense, but in the client-patron relationship, that's very European. There are those who dispense and those who receive. And that's kind of mixed in at times with a very Southern populist tradition. So you get the two of those together and you get the people who dispense—they can give you jobs, and there's just sort of no question that you become a client to their patronage. So they say, for example, "Those people from somewhere telling you that's no good [for the environment], well, they're just here starting trouble. Who gives you your paycheck?" And I think the region has that kind of trust in institutions, but in the sense of trust in institutions that have a visible representative.

I think there's a distrust in institutions that do not have visible representatives here, but a *great* trust in those institutions that have visible representatives. That's because it stands on this very European model of a village kind of patronage. If the man comes and says it to you, you believe it, but if it's paper that comes in the mail, forget it.

The fact that the industry did not cause the dissolution of the towns feeds that phenomenon. [Local people] still go to the same church, and the priest might be the brother of somebody that they know, and so on, such that the *model* has been allowed to continue and flourish. . . . So you have a real kind of European model of patronage, and *order,* to where you're not as apt to stand up and challenge things. . . .

[At the same time,] working in the oil industry became this sort of odd industrialized folkway, kind of weaved itself into a whole way of life. . . .

[Unlike] what industrialization has done to most other areas . . . it hasn't created tire factory cities like Akron, you know. . . .

This area was basically isolated for a long time . . . we didn't have bridges across the Mississippi until the twentieth century. . . . Whereas in California, virtually everyone who lives on the beach moved there from somewhere else . . . here, the only time you went to another place was to go work on an oil rig . . . [where the other workers] all knew each other. . . . [So] the only new people you met were the guys who were actually going out on the rigs, but who were maybe just about like all the other guys you'd always known.

In most of the coastal regions of the United States today, the situation is just the opposite. The coastal regions have seen extensive in-migration, often including a high proportion of persons who are attracted to those regions by a desire for high environmental quality. While some of the in-migrants are likely to have had at least some connection to the oil and gas industries—and in fact, as already noted, one of the California residents interviewed for this study was just such a person—there tend to be so few of them that they are more likely to evolve toward the nature-appreciative orientations of the majority of their neighbors than to convert those majority views. This tendency is especially likely in areas such as northern California, where there is no current base of employment that is directly dependent on local oil and gas development.

In addition, the California interviews suggested that interactions with oil industry representatives seem to have reinforced a sense of antipathy, rather than of social connectedness, with the oil industry. An unusually high proportion of the Californians reported that they had found industry and MMS representatives to be "arrogant," "pushy," or "not interested in what *we* think—they're only interested in preaching to us about how wonderful oil development would be," to quote just some of the characterizations that came up in the interviews. To note the obvious, these sorts of experiences are far less likely than are more agreeable interactions to lead to a climate of increased support.

Given that both of us have generally found industry and MMS personnel to be far more congenial than these descriptions would indicate, part of the explanation may have to do not with the "real personalities" of the people involved, but with the degree to which OCS issues have become politicized and polarized. At least since the time of Coleman's classic study of community controversies (1957), it has been known that the patterns of social interaction in a controversy can lead to a phenomenon that the two of us (Freudenburg and Gramling 1993, 358) have called a "spiral of stereotypes": When persons on different sides of an issue stop talking to one another (and worse, persist in talking *about* one another, as in characterizing the concerns of the other side as being ill-informed, self-serving, or irrational), the net effect can be to increase further the amount of polarization that was already present (see also Freudenburg and Pastor 1992).

As one simple indicator of the amount of polarization that now exists in northern California, it is useful to consider the frequent industry claims about the degree to which California opposition "just" reflects ignorance, selfishness, or irrationality, rather than reflecting legitimate concerns on the part of sensible citizens. These, clearly, are not the kinds of descriptions that seem to indicate an eagerness to understand one's opponents better. The California opponents of offshore drilling, on the other hand, seem to have equally little interest in getting to know better the representatives of the oil industry. As just one indicator of the extent to which the antipathy has become mutual, it is useful to realize that when Louisiana citizens think of oil industry representatives, they are thinking of their friends and neighbors, while in the words of one Californian, "When I think of the oil industry, I think of fat, pushy Texans in pointy-toed boots."

Overadaptation

Another way to understand the compatibility of Louisiana as a context for OCS development is to examine the degree of adaptation that has taken place. There is no question that social conditions in Louisiana have adapted significantly and often effectively to the needs of the oil and gas industry. As is the case for other species, however, human adaptations can have consequences of their own, and there can be reasons for concern if effective adaptations to one set of circumstances can introduce vulnerability to others.

As suggested by the Louisiana interviews that were summarized in the previous chapter, this concern is a real one; in fact, the degree of vulnerability in Louisiana is so severe that the coastal regions of the state provide a textbook example of *overadaptation* (Freudenburg and Gramling 1992). The concept will be discussed in greater detail in the following chapter, but in essence, overadaptation involves such a process that occurs in social and economic systems as well as the physical ones, and it is particularly evident in rural communities that become heavily dependent on, or adapted to, large-scale extractive activities.

Louisiana's vulnerability to the world oil market is illustrated by figure 5, which plots annual data on the price of crude oil on the world market, the world rig count, and the total employment in Lafayette and St. Mary parishes, from 1970 to 1989. As noted earlier, the period of rapid growth, most notably between 1973 and 1982, brought an incredible boom to the coastal Louisiana economy (cf. Gramling and Brabant 1986). Although the national media accounts of the region's boom-time economy were exaggerated, the period was one of considerable prosperity; as Festinger (1957) suggested

FIGURE 5
World Oil Rig Count and Price/Barrel Compared to Employment in Two Parishes

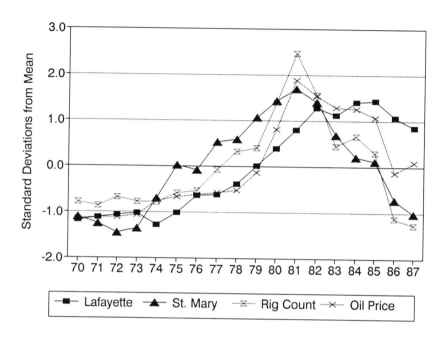

many years ago, it is difficult to be too critical in evaluating an activity from which one has benefitted—and upon which one remains dependent for any hopes of comparable, future prosperity. This, coupled with the industry's generally good record in terms of offshore environmental accidents, has tended to bolster the perception of offshore activity in the Gulf as involving relatively low levels of risk.

During the bust, however, the pain has become just as readily apparent, and one of the consequences of the volatility has been the emergence of an additional perceived risk in many parts of the country, namely a concern over risks to relatively stable ways of life. A major area of concern in northern California, as noted earlier, involved potential vulnerability to the vicissitudes of world commodity markets. Importantly, concerns were expressed not just about the "bust" periods, but also about the misleading "lessons" that might be taught by the ability to earn large amounts of money, quickly, during a boom. In the words of a fisherman who described an ambivalence about the highly paid clean-up activities after the *Exxon Valdez* spill, "Picking up dead animals for $17 an hour is no way to instill a value system in our children."

In addition, the concerns appear to have been related to what many of the people called the "artificial" nature of the boom conditions—as indicated perhaps most cogently by the observation that such times could not go on indefinitely. As any number of the Louisiana residents were able to attest, of course, such an observation has a very solid basis in fact (see also the more detailed analysis in Gramling and Freudenburg 1990).

The key differences between the Louisiana experience and experiences with OCS development elsewhere can be recapitulated as follows:

I. Historical Factors
   A. Early historical era—"age of exuberance"
   B. Temporal priority of development
   C. Incremental onset and evolution of industry
   D. Local origins and development of technology
II. Biophysical Factors
   A. Coastal topography
      1. Broad coastal marshes that preclude coastal highways and population concentrations
      2. Estuarine systems that include many potential harbors
      3. Low relief and energy levels
   B. Marine topography
      1. Broad, gradual slopes that increase area and decrease significance of conflicts with platforms
      2. Silt bottoms with few outcroppings and obstacles
III. Social Factors
   A. Low educational levels
   B. Extractive orientation toward biophysical environment
   C. Favorable patterns of contact with oil industry personnel
   D. Adaptability to oil development

The acceptance of the offshore oil and gas industry in Louisiana was aided by the specific historical circumstances under which the industry originated and grew. Acceptance has been further aided by the nature of the coastal and marine environments, including topography that has lowered the social salience of the coast, while simplifying access to offshore waters and reducing the likelihood of offshore conflicts with the fishing industry. The social factors of the human environment have further increased the acceptability of OCS activities, with the pre-development populations having been characterized by low educational levels and an extractive orientation toward the environment. Over time, these initially favorable conditions have been joined by the thorough (and favorably received) integration of oil-industry leaders and workers into the social fabric of the region—and by an extent of adaptation that helps to illustrate the region's very readiness to have welcomed oil development in the first place. Except to the extent that the factors in this table can be found in other places and times, unfortunately, the historical experiences in the central Gulf of Mexico may provide little in the way of useful guidance about what to expect in other OCS regions in the future.

5

Spinning and Slipping

Educating the Public

At a conference about offshore oil development in 1984, a Texas oil-industry official heard, apparently for the first time, the suggestion that it might be instructive for MMS to study the differences in perceptions of the oil industry in Louisiana and California. At an informal reception a few minutes later, he politely but confidently asserted that there was really no need for such a study. "In Louisiana, they know about oil, and they support it. In California, they just oppose it because they don't know enough about it. Those of us who are in the industry need to do a better job of educating people about the benefits, but once we do, there's no question the public will support it."

Since that time, the oil industry, and MMS, have spent years in "educating people about the benefits." To some extent, the effort would need to be described as successful, because the Californians we interviewed were quite well-informed about the benefits. They were also quite well-informed, however, about the drawbacks. So were their congressional representatives, who succeeded in imposing several congressional moratoria on California offshore leases. By 1990, after years of an "educational campaign" that could perhaps be characterized as having been successful in every way except for achieving its intended outcome, a different Texas oil man, George Bush, imposed a moratorium of his own.

At another informal reception at another conference about offshore oil development, this one in 1991, a different oil-industry representative, on learning that the two authors of this book were actually doing a comparison of OCS perceptions in Louisiana and California, offered helpfully that, in effect, he already knew the conclusion. He might as well have taken it directly from the Texan in 1984: "The more that people know about oil development," he said, "the more they support it. That's going to be true in California, too." When asked why opposition had continued to stiffen, even in the face of extensive efforts to "educate people about the benefits," his response continued to have the ring of familiarity: "We haven't done a good enough job of it yet. We just need to do a better job of getting the word out."

Perhaps—but perhaps not. A fundamental principle in the social sciences—just as in the biological and physical sciences—is that if the evidence

97

continues to contradict your hypotheses, then even if those hypotheses are among your all-time favorites, it is time to look for new ones.

As any educator can testify, "information" that is not believed is also not likely to make many converts. As the California interviews consistently revealed, the oil industry has a level of credibility along the northern California coast that is asymptotically approaching zero—but the Minerals Management Service has a level of credibility that is not much higher. Perhaps the reason most often mentioned in the interviews is that the agency is seen by residents as having little effective independence from the industry. As one Californian put it, "I've listened to a lot of people from the oil industry, and a lot of people from MMS, and usually, I can't even tell the difference." Another expressed his views in terms of a different kind of contest:

> I'm a real football fan, so let me explain it to you that way. When I go to a football game, and I see that somebody on the other side is cheating, or playing dirty, it doesn't really surprise me that much. That's sort of what you expect, if you know what I mean. But it's the referee's job to catch that kind of stuff. What really gets me irritated is when the *referee* isn't doing *his* job. In this case, the MMS is supposed to be the referee; they're supposed to be keeping everybody honest. But not only are they *not* keeping the industry honest—they're practically playing on the same team. They like to claim that they're wearing a different kind of uniform, but they're practically lining up on the other side of the field and helping the industry to carry the ball.

Concerns such as this one—concerns that a resource-management agency will fail to maintain a sufficient degree of independence from the industry that it is supposed to regulate—have been expressed before, and they have received a good deal of attention in academic circles. Perhaps most influential, over the long term, has been Bernstein's (1955) analysis of "institutional cycles," in which he argues that regulatory agencies and commissions frequently adopt the perspective of the regulated industry or group rather than the perspective of the "public good." Bernstein's basic argument is that even if regulatory bodies start out as robust enforcers, eventually the process may become so routinized that the primary mission becomes the maintenance of the status quo.

In the process, Bernstein argues, the position of the industry comes to dominate, and for two main reasons. The first is that the relatively greater organization of the industry can make it easier for the industry than for the broader public to identify interests and to organize political pressure (Olson 1965; see also Galanter 1974). The second reason is that there is more contact between the regulators and the industry (and hence more of a possibility for the shaping of shared ideals and values) than between regulators and the broader public (see also Friesema and Culhane 1976). As a result, the regulatory body tends to become increasingly isolated and insulated from the broader public. As the decision-making process becomes more insulated, agencies may informally "co-opt" powerful external forces (Selznick 1948), allowing certain groups to have access to the decision structure in return for cooperation, but such arrangements tend to offer little in the way of access for members of the general public who do not belong to the better-organized special interest groups.

More recent analyses have focused on a variety of additional considerations. At one extreme, perhaps the most straightforward argument involves the so-called "revolving door." According to this argument, many of the top-level people in federal agencies tend not to remain in their positions for extended periods of service. Instead, the regulated industry and the regulating agency are seen as regularly exchanging personnel, in a kind of revolving door between business and government (see Freitag 1975; Hershman 1977; *New York Times* 1975). This is a phenomenon that clearly does exist, although Freitag (1983) has argued that it is not as common as some have contended.

At the other extreme are arguments that are more subtle and complex. Stone (1980), for example, argues that agency actors are neither "captured" nor interested in future jobs, so much as they are alert to the fact that economically powerful actors can stop them from achieving their agencies' goals (and their own goals within the agency). As Stone puts it, "Business influence in particular controversies is not very impressive (business interests are often divided or defeated). . . . [B]usiness influence seems to be important in a way not measured by victories and losses. . . . The missing element . . . seems to be in the predispositions of public officials . . . . [who] over the long haul seem to favor upper-strata interests" (Stone 1980, 978). Stone argues that this "predisposition" results in part from the fact that government officials have economic constraints (they rarely have the revenues to accomplish all the goals they might wish to pursue), and in part from the importance of "associational considerations" (notably the fact that various groups in society have differential levels of ability to support or to block an official's efforts). In Stone's analysis, in other words, government officials do have goals and interests of their own, but they find that the most powerful groups in society tend to be the ones having the greatest power to aid in, or to frustrate, the achieving of those goals.

Concerns over the disproportionate influence of such powerful special-interest groups have led to a number of attempted reforms. It was partly because of concerns over agency susceptibility to organized industry pressure, for example, that the "new" regulatory agencies of the 1960s and 1970s, such as the Environmental Protection Agency, were established around broad tasks, such as "environmental protection," rather than being set up to regulate specific industries, as were older regulatory agencies such as the Food and Drug Administration or the Federal Communications Commission (for a broader discussion, see Caldwell 1982).

In recent years, however, a number of authors have raised broader criticisms that question the ability of any government agency, in a capitalistic system, to impose truly meaningful regulations. A number of issues have been raised (see particularly Offe 1984; Block 1987), but a key theme is that such agencies face a complex challenge—on the one hand, they are expected to protect the broader public, and on the other, they are often under intense pressure not to impose significant costs on regulated industries.

One obvious possibility for dealing with these potentially conflicting expectations is for the agency to seek a relatively cooperative relationship with regulated industries, particularly in cases where the agency has fewer resources for imposing regulations than the industries in question might be willing to invest in fighting back (see the discussion in West 1982). Such relationships might be expected to be particularly likely in the case of industries that are unusually large and powerful, and the petroleum industry is of course one of the largest and most powerful of all (cf. Sampson 1975; Engler 1961). Another possibility, as a number of observers have noted, is for an agency to emphasize actions that are high in symbolic value, but low in the actual, tangible costs they impose on industry (Freeman and Haveman 1972; Block 1987; Olson 1982).

For the purposes of understanding the issue of offshore oil, these discussions have two problems. One is that they generally fail to include much in the way of discussion about how such a delicate dance can be pulled off—a point to which we will soon return. The other is that, while it is certainly possible to see how some observers might conclude, as did the Californian quoted a few pages earlier, that "the referee" is playing for one of the opposing "sides," it appears to the two of us that there *are* important differences between the viewpoints in the industry and those in the MMS. In some cases, those differences have a subtlety that is difficult to discern without a fair amount of exposure to the issue, to be sure. In other cases, the spokespersons from *industry* seem to be the ones who take a position that is somewhere in the middle, between the views of MMS and those of the broader public—as for example in many of the earlier discussions involving the social and economic impacts of OCS developments in coastal communities, where at least some of MMS's top-level officials expressed adamant opposition to the recognition of local concerns, while a number of industry representatives were arguing for a more moderate approach. While there can be a tendency to dismiss such cases as somehow exceptional, involving for example "a class of ideologues" such as James Watt and the persons he brought to the top levels of the agency, what these examples do illustrate is the principle that problems can be created not just by a tendency for agency views to mirror the wishes of a regulated industry, but also by a tendency for an agency to develop a perspective that is uniquely its own.

Such a tendency, moreover, would be quite consistent with a common finding in the literature on organizations, which is that, by nature, organizations tend to develop more-or-less distinctive ways of viewing the world. In the view of Morgan (1986), in fact, "Organization rests in shared systems of meaning." As any number of authors have noted, effective organizations generally come over time to develop distinctive "cultures," or more accurately, "subcultures," of their own (see, e.g., Child 1981; Deal and Kennedy 1982; Frost et al. 1985; Handy 1979; Jay 1967; Jelienk et al. 1985; Killman et al. 1985; Morgan 1986; Pascale 1982; Schein 1982; Webber 1969); the implications of that point seem to us to be worthy of further attention.

The Persistence of Positions

Just as southern Louisiana and northern California tend to have developed relatively distinctive ways of viewing the world, in short, so too the subcultures that are developed in specific agencies and industrial sectors can help to define participants' views of "the way the world is." Such a tendency is not only understandable, but often useful—to note only the most obvious example, it can help agencies and industries to be more cohesive, and to interact more consistently with the world outside the subculture. The same tendency, however, can also create dangers, one of which is particularly obvious in the context of the interviews summarized above: As internal views become increasingly entrenched, agency and corporate personnel can start to encounter increasing difficulty in even understanding perspectives that are different from their own (Peters 1978; Smircich 1983). In the case of MMS, the review of the Environmental Studies Program by the National Academy of Sciences/ National Research Council (1992, 68) found the degree of distinctiveness to be worth noting explicitly:

> The Minerals Management Service must recognize that government officials have viewpoints—sometimes many viewpoints—and that there is no particular reason to expect the government view to be understood or even widely shared by others—especially if the others are far removed or culturally diverse. In other words, the officials' understandings and definitions of OCS issues and problems—and solutions to them—are as much based on those persons' perceptions, biases, culture, and experience as are those of any other affected person or community.

As should be clear by now, MMS and the oil industry tend to be quite comfortable when operating within the cultural contexts of the Gulf coast, where the offshore industry developed, but the failures of communication in

California tend to be truly world-class. This may not be just a coincidence. The subcultures of the agency and of the offshore industry, as noted throughout, have been shaped by the historical and geographic origins of OCS activities, which flourished in the supportive atmosphere of the Gulf region. As will be recalled, the first lease sales on the outer continental shelf, between 1954 and 1963, were all in the Gulf. It was not until 1966 that the first lease sale was held off southern California, and it was less than three years later, in early 1969, that Santa Barbara experienced the first major spill of OCS oil. Outside the Gulf, resistance seems to have been stiffening ever since; in the Gulf region, however, the agency and the industry have been developing a reasonably formidable working relationship, coming in the process to share a perspective that has remained virtually unchallenged for decades.

Under the circumstances, it is not difficult to understand why MMS and industry personnel would have come increasingly to think of the reactions from the Gulf region as being the ones that were "normal," no matter how distinctive or unusual the Gulf region may in fact have been. It may have been a matter of historical accident, but the region that was home to the original offshore development also happened to be one where the human environment was almost uniquely suitable for offshore oil and gas activities. Unfortunately, the ready acceptance of OCS development in Louisiana, both in earlier decades and in the present, may have led to insupportable expectations for the kinds of reactions that can be anticipated in other regions, whether now or in the foreseeable future.

The Science and the Spin Control

One obvious antidote to the problem of developing an excessively idiosyncratic perspective is to ask for independent advice, particularly scientific advice, and in many regards, the record of MMS in seeking independent advice would appear to be little short of exemplary. One of the provisions of the Outer Continental Shelf Lands Act, as amended (OCSLAA), involves an entire program of independent scientific studies, the Environmental Studies Program—officially and unofficially, "ESP." Even in comparison with the levels of research that have been carried out by other resource-management agencies to date, the ESP involves an impressive level of commitment, with roughly $500 million having been invested in scientific research over the past twenty years. From many perspectives, and certainly from the perspective of the agency, in an investment that already stands at roughly half of a *billion* dollars and that continues to grow, there is prima facie evidence of an exemplary search for independent insights. From other perspectives, however, the verdict is far less favorable.

At a meeting of the Scientific Advisory Committee for the Environmental Studies Program in 1988, the then-director of the program expressed what had become a substantial level of frustration: "We keep spending all this money on studies, and it never seems to do any good. Nobody ever believes the results." His analysis, as he noted at the time, was that the main cause of difficulties lay not with the studies, but with the public: "There are an awful lot of people out there who seem to be saying 'my mind's made up—don't try to confuse me with the facts.' " A similar view was suggested by a subsequent director of the entire Minerals Management Service some four years later, during a brief, informal discussion with several members of a new committee put together by National Academy of Sciences, this one a committee to review MMS plans for oil development off the northern coast of Alaska. He felt that MMS's studies had already demonstrated the safety of the proposed development, he explained; the studies didn't seem to him "to be the real issue." Instead, he continued, "It sometimes seems to me as though there are some people who will just be dead-set against leasing, no matter how safe it really is, and no matter how well it's managed."

These observations reflect views that are widespread within the Minerals Management Service. Persons from very different offices, across the entire agency, will often express the belief, usually with apparent conviction, that the agency is doing essentially everything that could be asked of it, but that public expectations and reactions are simply unreasonable. The common tendency is to accept such official statements at face value—as reflecting a perspective that is clear-headed, accurate, and balanced, as well as being "official." In this case, however, given the concerns that have been expressed about MMS—specifically including concerns about the adequacy of MMS's studies—it is necessary to examine the "official" views a bit more closely.

In particular, critics have long charged that while the agency may have succeeded in spending a good deal on its studies, it has shown little evidence of interest in the quality of those studies—only a desire for "conclusions that make the agency look good," as one critic put it during our interviews. The critics, of course, may not be free from bias in their views toward the agency, just as the agency may not be free from bias in its views toward its critics. We are fortunate, accordingly, that in this case we have access to a set of independent assessments of the Studies Program, provided by the National Academy of Sciences/National Research Council (NAS/NRC) at the request of the Minerals Management Service. On at least two occasions, the MMS has asked the NAS/NRC for an independent assessment of its program; the most recent assessment, which was requested in 1986, was given added impetus by the request from President Bush, noted in the opening pages of this book, for an assessment of the adequacy of the current scientific basis for decision-making.

The answers have not been flattering. A consistent set of conclusions from the various pieces of the formal review (National Research Council 1989, 1992, 1993) is that the agency's studies have had deficiencies in assessing the impacts on all three aspects of the environment—human, marine, and coastal. The deficiencies were found to be particularly striking with respect to studies of the human environment—so much so that the NAS/NRC concluded it would be necessary for MMS to start almost at ground zero. This point was driven home with quotations taken directly from the agency's enabling legislation:

> To balance the benefits of the leasing program with environmental risks, MMS must conduct studies that develop the information needed for "the assessment and management of environmental impacts on the human, marine, and coastal environments of the OCS and the coastal areas that may be affected by oil and gas development" (43 U.S.C. Sec. 1346 (a)(1)).
>
> MMS also must monitor the human, marine, and coastal environments of leased areas "in a manner designed to provide time-series and data-trend information which can be used for comparison with any previously collected data for the purpose of identifying any significant changes in the quality and productivity of such environments, for determining trends in the areas studied and monitored, and for designing experiments to identify the causes of such changes" (43 U.S.C. 1346 (b)). (National Research Council 1992, 14)
>
> With the exception of Alaska, the environmental studies program has not proved capable of collecting and analyzing the information needed for assessment and management of effects on the human environment. With the possible exception of the Santa Barbara County Monitoring Program in southern California—which is not an MMS initiative—MMS has no adequate program for collecting the information necessary to monitor changes subsequent to leasing. This is a particular fault for the Gulf of Mexico region.
>
> There is no quick fix to this situation. With the exception of the Alaska program and the monitoring program in southern California, there is not even much to build on. (National Research Council 1992, 64)

What the National Academy of Sciences assessment left unanswered was a further question—the question of why, after the expenditure of so much money, there could be so little to build on. While this question was not part of the charge to the NAS/NRC committee, we believe the question is an answerable one, as we attempt to demonstrate in this chapter. To understand the problem, however, it is useful to take a somewhat different perspective than has often characterized "academic" studies of policy issues in the past.

In general, although certainly not in all cases, academic debates over policy issues have tended to mirror the public debates in at least one important respect: most of these discussions tend in fact to have focused on overall or "policy issues." Relatively little attention, by contrast, has been devoted to actual steps that are required for policy implementation, which are generally seen as more mundane (cf. Pressman and Wildavsky 1973; but see Stone 1980).

Under the circumstances, it may be worth recalling the warnings of scholars such as Offe (1984) and Block (1987) about the temptation for agencies to emphasize their "toughness" in ways that are largely symbolic, all while doing very little in the way of taking tangible enforcement actions that would create actual costs for regulated industries. It is only a short distance from such warnings to an awareness that, despite all the attention that is devoted to "national policy," at least the stated policies may actually have less importance than they seem to have. At a minimum, there is the possibility that an agency might find it useful to develop overall statements of policies, goals, missions, and so forth, that respond well to public expectations, but that such broad statements might not correspond all that closely with an agency's actual behaviors. If questioned, the agency would be able to argue that "the details of implementation" should be left to its own discretion—an argument that is of course plausible, but also one that brings to mind the political saying that "the devil is in the details" (for a more detailed analysis, from which the following discussion is drawn, see Freudenburg and Gramling forthcoming).

It may also have something to do with the well-known admonition to "watch what they do, not what they say." The problem, in a nutshell, is that while an agency's public stance—"what they say"—may in fact suggest an appropriately vigorous approach to enforcement, such a stance may not guarantee that "what they do" will show the same degree of vigor (Stone 1980; but see also Clark 1968). In the present context, more specifically, this perspective would point to the possibility that a general weakness of enforcement vigor, at least in the eyes of the critics, might be combined with broad "policy statements" that are intended to convey just the opposite impression—being complete, for example, with an emphasis on "basing our decisions on science, not politics," and on "taking all reasonable steps to protect the environment." Broad policy statements such as these might or might not keep the agency from making its day-to-day decisions in ways that would reflect a general difference to the concerns of the agency's core, industrial clientele, such as desires for increased production and/or lower costs of operation.

While the possibility deserves closer examination, a warning is necessary: explanations such as these can tend to sound quite conspiratorial, while at least in our own view, the reality tends to be far less clear-cut. In most cases, the professional-level personnel, who make up the majority of agency employees, tend to be quite good at resisting overt conspiracies, direct pressures to falsify results, and other obvious sources of bias. The problems, we believe, are more likely to be serious when the pressures toward bias are less obvious and do *not* involve what most people would recognize as "conspiracies."

It is precisely for this reason that an agency subculture can have such an important influence even on scientific studies. The greater the degree to which a given way of viewing the world tends to be accepted as appropriate or "natural," the greater its potential power. Paradoxically, in short, the effects of an organizational subculture may be the most powerful precisely when they are most subtle—when they surround us, and are taken for granted, every bit as much as the air we breathe—or the questions we may not even think to ask.

The Statutes and the Slippage

Given the subtlety of such considerations, how might they be studied, and how might such studies be kept as objective as possible? We believe that the clearest test is to look for "bureaucratic slippage"—the tendency for broad policy statements to be successively reinterpreted, both over time and across multiple layers of regulatory implementation. The net result can resemble the childhood game where a "secret" is whispered to the first child, who then whispers it to the next, and so on; the eventual secret, or the eventual implementation of the policy, can prove to have very little resemblance to the statement that started the process (see Freudenburg and Gramling forthcoming for a more detailed discussion).

Given the foregoing discussion, there is a distinct possibility that the bureaucratic slippage will be anything but random. As noted, at least some authors have concluded that an agency's ultimate implementation of a law is likely to devote far more attention to the production-related concern of the agency's core industrial clientele than to broader concerns that the agency may see as secondary—perhaps even including the protection of the human, marine, and coastal environments. As a self-described "country lawyer" once put it in an informal conversation with one of the authors, "When an agency *really* gets me irritated, then I know it's time to do what I always do when they get me really irritated. That's my sign that it's finally time to go back and read the original statute."

To approach the same problem from the perspective of social scientists rather than lawyers, alternatively, one of the real risks of distinctive agency subcultures is that loyal employees of such organizations will come, over time, to forget that the distinctive views within the agency may or may not actually reflect "the will of the governed"—or even what, according to the law, the agency's "real goals" are expected to be. This is particularly so if the reward structure within the agency tends to reinforce such forgetfulness. If this is the case, the consequences can be expected to show up in differences between the laws and the agency reality—subtle differences in the case of the agency's broadest statements of goals and purposes, somewhat larger differences as we move away from the broadest overview statements and toward the mid-level guidelines for agency performance, and quite clear differences, finally, between what the law "says" and what the agency actually "does," in the day-to-day performance of its duties.

What the law says, in this case, is easy enough to discern. The Outer Continental Shelf Lands Act, as amended (OCSLAA), provides a relatively concise, coherent, and readily understood statement of national policies toward the resources of the OCS. The act lists six overall principles. The first is that, as reflected in the decisions growing out of the Tidelands cases discussed in chapter 2, offshore resources are the property of the United States as a whole. The second is that the act is not intended to restrict offshore rights to navigation and fishing on the high seas. The last is that it is important for development to be done in a way that is both safe and sensible. The central three points, however, are worth examining in their entirety, to avoid the potential for any subtle changes of meaning that might result from paraphrasing; those are the goals that are reported in the top half of table 4.

As we have noted, the most reasonable expectation is for the degree of slippage to be only modest in cases where an agency is merely restating its broad goals, and that does seem to be true in the case of MMS. By the time the Minerals Management Service published a "Mission Statement" in 1987, the statement (1987a, vol. 3, appendix F: 1–2) included a version of goals that it described as having been "stated in the OCS Lands Act Amendments (OCSLAA) of 1978 (P.L. 95-372)" (although the Mission Statement does not refer to specific sections of the act). The restatement, which has been reproduced as the bottom panel of table 4, shows linguistic shifts that are relatively subtle, for the most part, yet the shifts may be instructive.

The most obvious difference is that the four points in the Mission Statement have no clear correspondence with the six main points listed under the heading of "Congressional Declaration of Policy" (43 U.S.C. 1332). In addition, there is at least a shift in emphasis. In the agency's restatement, objectives 1, 3, and 4 all have to do, in essence, with economic and developmental considerations. Only objective 2 now refers to protecting the human, marine, and coastal environments, and two further changes have taken place even in the statement of this objective. First, rather than seeing a requirement that the MMS "protect" the environment, the agency sees only a requirement to "provide for" protection of the environment. Second, while the OCSLAA calls for "expeditious and orderly development, *subject to environmental safeguards*" (emphasis added), the agency's restatement calls for only that degree of environmental protection that is "concomitant with mineral resource development." Such a linguistic shift may be a mere coincidence, of course, but it also brings to mind one of the consistent complaints of northern California residents—namely that the agency was only willing to

TABLE 4
Agency Goals: Statutory Language and Agency Restatement

Statement of Goals, OCSLAA

It is hereby declared to be the policy of the United States that . . .

(3) the outer Continental Shelf is a vital national resource reserve held by the Federal Government for the public, which should be made available for expeditious and orderly development, subject to environmental safeguards, in a manner which is consistent with the maintenance of competition and other national needs;

(4) Since exploration, development, and production of the minerals of the outer Continental Shelf will have significant impacts on coastal and non-coastal areas of the coastal States, and on other affected States, and, in recognition of the national interest in the effective management of the marine, coastal, and human environments—

(A) such States and their affected local governments may require assistance in protecting their coastal zones and other affected areas from any temporary or permanent adverse effects of such impacts; and

(B) such States, and through such States, affected local governments, are entitled to an opportunity to participate, to the extent consistent with the national interest, in the policy and planning decisions made by the Federal Government relating to exploration for, and development and production of, minerals of the outer Continental Shelf;

(5) the rights and responsibilities of all States and, where appropriate, local governments, to preserve and protect their marine, human, and coastal environments through such means as regulation of land, air, and water uses, of safety, and of related development and activity should be considered and recognized (43 U.S.C. 1332).

Restatement of Goals, MMS "Mission Statement" (U.S. Minerals Management Service 1987):

The four major goals for the comprehensive management of OCS minerals are

1. To ensure orderly development of the marine mineral resources to meet the energy demands of the Nation.

2. To provide for protection of the human, marine and coastal environments concomitant with mineral resource development.

3. To provide for receipt of a fair market value for the leased mineral resources.

4. To preserve and maintain free enterprise competition.

impose the kinds of environmental regulations that would do nothing to slow the pace of oil development—particularly given that the basic dictionary definition of "concomitant" is something along the lines of "accompanying" or "occurring concurrently."[11]

A similar degree of slippage seems to have taken place in interpreting—and especially, *implementing*—the goals of the Environmental Studies Program. The OCSLAA provisions and the agency restatements are reported in table 5. Again here, by the time the agency has restated the objectives, a certain amount of slippage has taken place, but it is relatively minor; the more significant slippage takes place in the steps that are closer to actual implementation.

Perhaps the major form of slippage in the restatement of objectives comes from the apparently minor fact that the agency's restatement is far less specific than the requirements in the act itself. While such a lack of specificity can serve many functions, including mere simplification, it has the potential to offer one kind of contribution that an assessment of bureaucratic slippage cannot afford to overlook: with such a vague restatement, it would be possible for the specific "implementation" guidelines to be entirely consistent with the restatement, even if they *fail* to be consistent with the law's more explicit requirements.

At a more detailed level of analysis, two other points are worthy of attention. The first point is a subtle one, but it has an importance that will become apparent when we examine the specific implementation guidelines. The law (43 U.S.C. 1346(b)) explicitly requires that studies be done *after* leases are issued, and not just beforehand: "*subsequent to* the leasing and developing of any area or region, the Secretary shall conduct . . . additional studies" (emphasis added), the purposes of which are to include the documentation of the impacts that are actually created by leasing. The restatement calls more vaguely for "information on the status of the environment upon which the prediction of impacts of OCS oil and gas development may be based" and for the agency to "provide a basis for future monitoring of OCS operations." As will be seen below, moreover, there is very little evidence that the agency's "information on the status of the environment" has actually included much in the way of scientific information on the status of the environment "subsequent to" leasing decisions. The second point is one that may be minor, but that may also have relevance in understanding the degree to which an agency has developed views of the world that are distinctive or insular. There is no mention of the requirement that "the Secretary" (and hence the agency) should "plan and carry out such [studies-related] duties in full cooperation with affected States," and there is no mention of the option of using information from other federal agencies that have prepared environmental impact statements, are conducting studies, or are monitoring the environment.

While such a restatement can provide an important degree of insulation between the law's requirements and the agency's actual actions, the document that plays the key role in determining which studies are supported is not the overall statement of policy goals, but rather the list of specific implementation criteria—in this case, a set of criteria developed jointly between MMS and the Office of Management and Budget (OMB). This document is even

TABLE 5
Studies Program Goals: Statutory Language and Agency Restatement

Goals of Environmental Studies Program, as provided by OCSLAA

(a) (1) The Secretary shall conduct a study of any area or region included in any oil and gas lease sale in order to establish information needed for assessment and management of environmental impacts on the human, marine, and coastal environments of the outer Continental Shelf and the coastal areas which may be affected by oil and gas development in such area or region . . .

(2) Each study . . . shall be commenced . . . not later than six months prior to the holding of a lease sale. . . .

(3) In addition to developing environmental information, any study of an area of region, to the extent practicable, shall be designed to predict impacts on the marine biota which may result from chronic low level pollution or large spills associated with outer Continental Shelf production, from the introduction of drill cuttings and drilling muds in the area, and from the laying of pipe to serve the offshore production area, and the impacts of development offshore on the affected and coastal areas.

(b) Subsequent to the leasing and developing of any area or region, the Secretary shall conduct such additional studies to establish environmental information as he deems necessary and shall monitor the human, marine, and coastal environments of such area or region in a manner designed to provide time-series and data trend information which can be used for comparison with any previously collected data for the purpose of identifying any significant changes in the quality and productivity of such environments, for establishing trends in the areas studied and monitored, and for designing experiments to identify the causes of such changes.

(c) The Secretary shall, by regulation, establish procedures for carrying out his duties under this section, and shall plan and carry out such duties in full cooperation with affected States. To the extent that other Federal agencies have prepared environmental impact statements, are conducting studies, or are monitoring the affected human, marine, or coastal environment, the Secretary may utilize the information derived therefrom in lieu of directly conducting such activities. . . .

(d) The Secretary shall consider available relevant environmental information in making decisions (including those relating to exploration plans, drilling permits, and development and production plans), in developing appropriate regulations and lease conditions, and in issuing operating orders.

Restatement of Environmental Studies Program Goals, MMS "Mission Statement"

1. Provide information on the status of the environment upon which the prediction of impacts of OCS oil and gas development may be based.

2. Provide information on the ways and extent that OCS development can potentially impact the human, marine, biological, and coastal environments.

TABLE 5
(*continued*)

3. Ensure that information already available or being collected under the program is in a form that can be used in the decision making process associated with a specific leasing action or with the longer term OCS mineral management responsibilities.
4. Provide a basis for future monitoring of OCS operations (U.S. Minerals Management Service 1987a, vol. 3, appendix F: 1–2).

more dense with mind-numbing bureaucratic details than are the act and the MMS's restatements of the same—so much so that many who try to read it find their eyes glazing over well before they comprehend the minutia of meanings. This set of criteria, however, rather than the law or any broad statements of agency policy, is what provides the official basis for actual agency decisions about which studies to support or not to support, and thus they are worthy of closer attention.

While several sets of criteria were originally put forward, two sets of criteria are particularly important. The "A" criteria have to do with the specificity of a legal requirement for doing the study, and the "B" criteria have to do with the agency's sense of urgency with respect to the study. Note in particular that the "A" criteria emphasize the kinds of priority decisions that would be made by courts, lawyers, and politicians, while placing far less emphasis on the reasonable-sounding but broad considerations such as "establish[ing] information needed for assessment and management of environmental impacts on the human, marine, and coastal environments," which are emphasized both by most scientists and by the policy statements in the law itself (see table 6).

The net result of this approach to "implementation" carries more than a little irony. The law provides a mandate for the program that is broad and inclusive—to establish the information necessary to assess and manage the impacts of oil and gas activities on the OCS, including the collection of information over time, and to identify trends and causes of trends. Such broad considerations, however, have disappeared entirely by the time we get to the decision-making criteria actually followed by the agency. The "A" criteria give the highest priority to information that is "essential" not because of any (other) logical reason, but because the information has been "ordered by a court," because it is "explicitly" required by a federal statute, or because it is similar to a study that was "required by previous agreement for settlement of litigation"—provided that the requirement "came from a similar OCS region." The "B" criteria give highest priority to studies that "must be initiated" immediately, "to be completed in time for use in a specific leasing . . . decision."

TABLE 6

Agency Criteria for Selecting Studies

"A" Criteria: Mandate for Conducting Study

1. Information for a study is essential for a specific leasing, lease management, or program management decision because the study is:

   a. Ordered by a court to support a sale or program specific decision; or

   b. Explicitly required by existing Federal or State statute or by an Agency directive.

2. The study will provide critical information for a specific leasing or lease management decision involving environmental risk or impact or for a specific program management decision, or a similar study is required by previous agreement for settlement of litigation in a similar OCS region.

3. The study will provide useful information for a specific leasing or lease management decision involving environmental risk or impact for a specific program management decision.

4. The study information will not affect a leasing or a lease management decision involving environmental risk or impact or a specific program management decision. However, it could contribute to improved leasing, lease management, or program management decisions through the enhancement and/or refinement of the quality of the database.

"B" Criteria: Timing and Content for the Required Information

1. The study must be initiated within the budget period at issue:

   a. In order to be completed in time for use in a specific leasing, lease management, or specific program management decision; or

   b. Because the study is a necessary prerequisite for another study to support a leasing, lease management, or program management decision; or

   c. Because the study is a continuation or logical extension or a consequence to an ongoing study to support a leasing, lease management, or program management decision.

2. A study can be deferred until the next budget period; however, such a deferral can create a significant risk that the study cannot be completed in time for the use in immediate leasing, lease management, or program management decision making.

3. A study can be deferred until the next budget period and can still be completed in time for use in forthcoming leasing, lease management, or program management decision making.

Class 1—A1B1	Class 7—A1B3
Class 2—A2B1	Class 8—A2B3
Class 3—A1B2	Class 9—A3B3
Class 4—A2B2	Class 10—A4B1
Class 5—A3B1	Class 11—A4B2
Class 6—A3B2	Class 12—A4B3

It is only in the fourth category of the "A" criteria that we see any language remotely pertaining to time-series analysis, or documentation of trends over time; that language comes in a scarcely ringing endorsement of information that "will not affect a leasing or lease management decision . . . [but] could contribute to improved . . . decisions through the enhancement and/or refinement of the quality of the database." There is nothing in criteria "B" that would even hint at such a consideration.

The triumph of legal considerations over scientific ones becomes all the more clear in considering the agency's twelve "classes" of studies, a form of prioritization that results from combining the A and B criteria, as listed at the bottom of table 6. In practical terms, the central consideration has to do with the "cutoff level" for funding decisions; for years, the level of funding has consistently proved to be insufficient to support any studies below "class 2." The "A4" criterion, which provides at least an indirect and lukewarm endorsement of the very kinds of studies called for in 43 U.S.C. 1346(b), does not show up until *class 10*. In effect, while the sensible management of impacts, over time, may be supported in the abstract—critics might refer to such statements as "lip service"—the agency's official implementation guidelines effectively make it almost illegal for the Environmental Studies Program to support some of the very kinds of studies that are explicitly called for in the law itself.

And yet that is of course not the end of the potential for bureaucratic slippage. Further redefinition can take place in the necessarily judgmental processes that lead up to the agency's decisions about just which studies are "explicitly" required, "immediate" in their urgency, and "necessary" for decision-making. These internal agency judgments are even less accessible, analytically, than are the funding criteria, given that the final decisions tend to be made behind closed doors and to be reported in passive voice ("the decision has been made to support the following studies").

The net results of the slippage are perhaps easiest to quantify in terms of the differing levels of research that have actually been done in the various Outer Continental Shelf regions. As should be clear by now, the Gulf of Mexico is the site of virtually all of the nation's OCS production and of the heaviest concentration of OCS impacts; as of the time of this writing in 1992, there were approximately 3,800 offshore production platforms in the Gulf, connected by thousands of miles of under-sea pipelines, and maintained by a massive fleet of supply vessels. On the entire remainder of the nation's Outer Continental Shelf, there were exactly 24 production platforms, all of them concentrated in the Santa Barbara Channel of MMS's Pacific Region—with no OCS production at all being found in the agency's other two regions, Alaska and the Atlantic (Gould et al. 1991). Over the entire history of the Environmental Studies Program up to October of 1990, however, only

15.17% of the funding to address the effects of Outer Continental Shelf oil and gas activities had been devoted to the Gulf of Mexico—a lower level than for any other MMS region.

The results also showed up in what for many years were the agency's policies toward impacts on the human environment—informal but reinforced by positions taken in litigation—that entire classes of impacts deserved no research at all. In northern California, for example, when the agency's own Scientific Committee repeatedly noted the need to deal with the impacts already being created, the consistent response from the agency officials, in essence, was that there could not *be* any impacts until physical activities began offshore. Scientifically, such an agency position is difficult to defend. As noted in the evaluation done for the National Academy of Sciences/National Research Council (1993:3), "Humans and their social systems can and do respond to information [itself]. . . . Alterations in social systems—*before any biological or physical change has occurred* . . . are real, and so are their consequences" (National Research Council 1993, 3, emphasis in original). At the time when one of us first began his service on the MMS Scientific Committee, however, the director of the Environmental Studies Program not only ignored this fact but went even further, arguing in essence that it was illegal for the agency to spend any money in doing studies within three miles of the coast, except in Alaska, where additional laws, and "the federal government's historic trust relationships with Alaska natives," created different and more specific obligations. A straightforward reading of the law, of course, shows that at least two of the three kinds of studies that the Environmental Studies Program was explicitly designed to perform—those involving the human and coastal environments—would seem to require attention to variables that are found well within three miles of shore.

When this discrepancy was first pointed out to agency personnel, at a subsequent meeting of the Scientific Advisory Committee, the immediate reaction included a great deal of agitated mumbling. At the next day's meeting of the same committee, however, after an evening that appeared to have involved some equally agitated consultation, the official position had changed a bit. Everybody knew, the director pointed out, that the Environmental Studies Program was designed to assess likely effects on the human, marine, and coastal environments. The problem was not in the absence of broad, legal authority or permission for doing such studies, but in the agency's more pragmatic determination "to be careful stewards of limited public funds." The director's position, as restated or at least clarified, was that there was no legal basis for the agency to impose limitations on what companies could or could not do within three miles of shore, and it was for that reason he "would have difficulty justifying expenditures for studies on the human environment." He had no interest, he continued, in spending scarce agency funds on studies unless he could be shown in advance that the agency would be able to make practical use of the results.

Even at this point, it would have been difficult to accuse this official of overreaching the authority that was granted to the agency by the law. Among the many provisions to which he might have referred, for example, OCSLAA states that regulations for the administration of leasing "shall include, but not be limited to," a set of provisions that range up to "cancellation of a lease or permit" in cases where the agency decides that "continued activity" would "probably cause serious damage . . . to the marine, coastal, or human environment" (43 U.S.C. 1334 (a)). As noted above, moreover, perhaps the most important single factor governing the magnitude of development-stage socioeconomic impacts is the speed of leasing sought by the MMS—a factor that, to most points of view, would seem to be well within the agency's control.

In fairness, however, several points should be granted. The first is that, to the extent to which the decisions over the rate of leasing are political, rather than technical, the leasing-rate factor may not actually be within the control of at least the agency's technical-level personnel. The agency's leaders, as noted earlier, tend to be quite persistent in claiming that their decisions are *not* influential by political considerations, but if such claims were literally true, they would be little short of remarkable. Even at somewhat lower levels in the agency, in positions that are generally expected to be "more technical" and less political—and indeed, even in the case of a program that is supposed to be scientific—the director of the program may find that technical decisions are constrained by higher-level decisions that are political.

The second point is that social scientists need to accept some of the responsibility for the lack of progress. At least until recently, relatively few social scientists have provided much in the way of guidance for the agency about how to go about the task of studying the human environment. To some extent, of course, this has been a chicken-and-egg problem, in that the agency has failed to support the work that would have provided the necessary foundation, and there has been little reason to believe that the guidance would have been taken seriously even if it had been offered. Even so, responsibilities need to be acknowledged, and in this case, it is the responsibility of the social science community to suggest a better set of alternatives. It is for this reason that, in the following chapter, we will provide at least an outline of an approach that we feel could offer the agency much more progress in its efforts to understand "impacts on the human environment."

Third and finally, as noted earlier, the agency's views on the human environment have finally begun to evolve toward a less creative reading of the law. The gentleman quoted above is no longer the director of the Environmental Studies Program. Even before he left that position—and before the National Academy of Sciences produced its critical assessment of the agency's failure to make integrated use of social science research—the Studies Program was beginning to respond to similar assessments from its own Scientific Advisory Committee. As one small if tangible piece of evidence of the changes taking place in MMS, it is worth repeating that this book is one of the results of a social science study that was funded by the same director's office.

To date, however, the rate of progress has continued to be glacial. The Alaska region has devoted a significant amount of attention to studies of social and economic impacts. Outside of that region, however, rather than receiving something like one-third of the attention and funding from the Environmental Studies Program, "the human environment" has received about the same amount of attention as was devoted to non-economic aspects of the human environment in the draft environmental impact statement for the proposed northern California oil lease sales—not 33%, and in fact not even 3.3%, but a level even lower than that.

6

A Framework for the Future

If the Minerals Management Service is to continue its current progress toward understanding the human environment, the agency will need a much more explicit plan to guide its course of study. As is probably clear to anyone who has ever held responsibility for trying to implement a policy, it is far easier to criticize the performance of the past than to offer constructive suggestions for how to do things better in the future. Given that both of us have long worked with offshore development issues, and that we count numerous MMS employees as friends and colleagues, we believe we need to go beyond identification of problems, to provide at least a set of suggestions for how the current situation might be improved.

As part of that effort, this chapter summarizes a framework that appears to offer a relatively straightforward way of conceptualizing the fuller range of impacts on the human environment. While the task is complicated by the wide range of experiences associated with offshore developments—from the so-called "pre-development" impacts in California, through the much longer-term impacts in Louisiana—the suggested framework reflects the fact that the very breadth of experience can also offer important advantages. Particularly in the aggregate, OCS activities can provide useful illustrations of the potential range of "impacts on the human environment," both across time and across the different components or systems of the human environment (for articles that discuss the implications of this framework for the social science community, see Freudenburg and Gramling 1992; Gramling and Freudenburg 1992a).

A starting point for our discussion is that, if future studies are to deal adequately with impacts on the human environment, those studies will need to reflect the fact that development-related impacts follow channels that are *social as well as physical:* humans and social systems adapt not just to the physical and demographic disturbances that have received the majority of attention to date, but also to the social construction of opportunities and threats—a process that will be discussed more fully below—and increasingly over time, to the accumulating consequences of earlier adaptations and experiences.

There is an important degree of correspondence between the stage of development and the channel of influence. While the degree of correspondence

117

Louisiana coastal hotel. This is one of the few hotels on Grand Isle, the only Louisiana barrier island with road access, and for all practical purposes, the only beach-front development in the state of Louisiana.

California coastal hotel. While this is just one of the many California coastal hotels, this hotel complex alone has roughly the same number of rooms as do all of Louisiana's coastal hotels and motels, combined.

is significantly less than one-to-one, some of the earliest (and in certain cases, some of the most severe) impacts of development are likely to be associated with the social construction of opportunities and threats, and many of the most severe longer-term impacts result from the accumulation of experience over time. Our discussion will begin, accordingly, with the earliest, or opportunity-threat, impacts.

The Initial or Opportunity-Threat Phase of Development

In the physical or biological sciences, it may in fact be true that no impacts take place until a project leads to concrete alterations of physical or biological conditions. In the case of the human environment, by contrast, observable and measurable impacts can take place as soon as there are changes in *social* conditions—which often means from the time of the earliest announcements or rumors about a project. As we have noted elsewhere,

> Speculators buy property, politicians maneuver for position, interest groups form or redirect their energies, stresses mount, and a variety of other social and economic impacts take place, particularly in the case of facilities that are large, controversial, risky, or otherwise out of the range of ordinary experiences for the local community. These changes have sometimes been called "pre-development" or "anticipatory" impacts, but they are far more real and measurable than such terminology might imply. Even the earliest acts of speculators, for example, can drive up the *real* costs of real estate (Freudenburg and Gramling 1992, 941).

In the interest of greater accuracy, our technical articles (see also Gramling and Freudenburg 1992a) refer to this early stage as the "opportunity-threat phase" of development. The terminology reflects the fact that social and economic impacts characterizing this phase derive predominantly from the efforts of interested parties to identify, to define socially, and to respond to the anticipated as well as the ongoing implications of development— whether as "opportunities" (to those who see the changes as positive) and/or as "threats" (to those who feel otherwise). For purposes of non-technical discussion, these early impacts can also be discussed simply as "planning phase" impacts. The essential point to realize is that *the process of negotiating the "real meanings" of development may play a key role in determining the social and economic impacts of the facility or activity.* This is true even in cases where no facility is ultimately constructed, and also in cases where construction and development do take place, although additional impacts can be expected from either the cessation or the intensification of activity, as will be noted further below.

Clearly, the definitions of opportunities and threats are shaped by a community's prior experience and present interests. This is one of the reasons why proposals for offshore oil development along the coasts of Florida and northern California, for example, have been so contentious that they eventually required Presidential intervention, while until recently, similar proposals for areas off the coasts of Texas and Louisiana have inspired consistent local backing. As noted above, the support activities that are necessary to construct, transport, and maintain oil platforms, and to drill the exploratory and production wells, have already transformed the coastal and human environments along the Gulf of Mexico (cf. Gramling and Brabant 1986; Gramling and Freudenburg 1990); today, thousands of people in Texas and Louisiana are directly dependent on offshore oil for their employment, and this fact clearly contributes to a tendency, even for those who are not oil employees, to see offshore oil in terms of opportunities. For current residents of Florida or northern California, by contrast, there is little reason to think of OCS development as offering opportunities for maintaining current jobs, either for their neighbors or for themselves, and as noted above, the proposals for OCS developments in these regions have been made during an era when environmental sensitivities have become more intense, bringing increased salience to the potential threats.

A closer look reveals, however, that "experiences" and "interests" are insufficient, by themselves, to explain the impacts that emerge. The processes of identifying and defining opportunities and threats, far from being automatic or self-regulating, are inherently social and can often be contentious. As Kunreuther and his colleagues (1982), Molotch (1970), Stallings (1990), and others have pointed out, any number of participants can be involved in the process of attempting to influence the socially negotiated definitions of development: project proponents, local influentials, the media, affected citizens, various levels of government agencies, environmental and developmental groups, and more. The impacts that emerge are shaped by the characteristics of the negotiation process, such as fairness and openness (Creighton 1980; Howell et al. 1981), by the definitions of the situation that are propounded and ultimately accepted, and increasingly over time, by the actions taken by individuals and groups, in response to their definitions of the situation, to each other, and to the negotiation process itself.

To date, most of the participants in debates over OCS development have argued whether the development is "best" seen as offering opportunities or threats; for the most part, these arguments *inherently* involve an emphasis on selected implications of development, with other implications being denied or overlooked. In many cases, the debate is made more complex because claims of opportunities focus on one or more of the systems of the human environment (most often economic systems) while claims of threats focus on other systems (most often biophysical/health and/or social systems). A more fruitful approach for the future, by contrast, is to realize that the very *debate* over the opportunities and threats is a source of conflict, difficulties, and socioeconomic impacts.

While there is a degree of overlap, the opportunities and threats can be seen as having implications for six systems of the human environment. The following discussion will focus briefly on each.

Biophysical/Health Systems. To begin with the most obvious system of the human environment, the kinds of facilities that are seen by many persons as involving unacceptable risks, such as toxic waste facilities in some communities or offshore oil developments in others, are often the ones that generate concerns about potential threats to the biophysical environment and/or community health. With the exception of concerns such as the potential human health impacts of air pollution off the California coast, most of the more contentious biophysical/health impacts of proposed OCS developments have to do not so much with the potential for *human* death or health damage, but with threats to the components of the biophysical environment that humans value. (Even the *Exxon Valdez* spill, for example, has not been blamed for near-term human deaths.)

As suggested by the interviews that were summarized in chapters 3 and 4, OCS-related concerns can range from the concrete (as in the potential for habitat degradation for fish or marine mammals) to the symbolic or abstract (as in the potential for industrial developments that would be incompatible with the aesthetic characteristics or cultural or religious significance of particular areas). To characterize the latter concerns as "merely symbolic," however, would be to misrepresent them in a serious way—as well as to misunderstand their intensity and significance. These concerns are real, constitute real impacts in their own right, and lead in turn to real action, organization, and political activity. The chanting and cheering of thousands of OCS opponents at the 1988 hearings in Fort Bragg, California, for example, was not just imagined by the bureaucrats at the front of the room. Like the organized resistance at Key West, Florida, in 1989, where President Bush's task force was conducting hearings, such an outpouring of emotion provides visible evidence that OCS activities have already altered people's lives in both regions, even though no offshore lands had been leased in advance of either hearing. So obvious were the "real" characteristics of these concerns, in fact, that some of the members of the presidential task force were visibly frightened by the intensity of the reactions.

Less commonly recognized is the fact that proposed developments can also be interpreted as opportunities for improvements to the environment or public health. Under favorable circumstances, community leaders and planners may find the development to offer an opportunity they have always wanted—to use "growth management planning" as a tool to develop a more attractive community, for example, or to develop new recreational facilities and to "get rid of that old eyesore downtown." Local leaders may even see the development as a way to improve community health by upgrading inadequate water or sanitation facilities, attracting new doctors to understaffed medical facilities, or otherwise upgrading the facilities, services, and environmental amenities of the region. During the period of most rapid growth of offshore activities in coastal Louisiana (1974–81), future development was used as the primary vehicle for obtaining federal assistance to upgrade inadequate water and sanitation facilities, and at least in one case to build a new hospital (cf. Gramling and Joubert 1977).

Cultural Systems. In many cases, the same developments that are seen as posing threats to physical, economic and/or social well-being will almost necessarily imply threats as well in the realm of culture and norms. These threats are most obvious in cases of indigenous or native cultures, whose very survival may be threatened by development. Disruptions may be created not only by potential technological accidents such as the *Exxon Valdez* oil spill and by dramatic increases in contact with large numbers of outsiders, but also simply by the kinds of increased dependence on money economies that can threaten subsistence activities. Given that, in many cases, it is through subsistence activities that many of the norms and customs of a culture are passed from generation to generation—and that the culture dies unless it is passed on to a new generation—the potential for disruption is considerable.

Even in the case of so-called "mainstream" cultures, however, proposals for new developments and the ensuing battles can threaten citizens' views of how the world "ought to" work, particularly with respect to the degree of congruence between expected and actual behaviors of governmental and other authorities. There is a good deal of discussion in the open literature that reveals clear parallels to the comments made by many coastal residents interviewed for this study, such as the man who became concerned when the governmental "referee" seemed to start playing on the "side" of the oil companies; elsewhere, as well, even previously neutral citizens will express often intense levels of anger about the failures of the government officials to exhibit appropriately neutral behaviors. Levine (1982), for example, refers to the increasing incredulity of the Love Canal residents who found governmental authorities to be unresponsive to their plight, often seeming to show greater concern about the health of the agency's budget than of the community's residents, and Krauss (1987) provides a case study of the "progressive radicalization" of a worker whose initial faith in the proper conduct of authorities is progressively replaced by skepticism, mistrust, and even cynicism (see also Finsterbusch 1988; Edelstein 1988; Clarke 1988a; Freudenburg 1993).

As Erikson (1976) notes, one of the functions of a sociocultural system may be to keep certain disturbing questions unasked or unthinkable, and one of the consequences of a disaster may be to cause people to question what previously had been taken for granted; Molotch (1970) raises a similar point in connection with the 1968 oil spill in Santa Barbara. What we are suggesting here is that the very actions taken in the process of defining a facility's potential impacts as opportunities or threats (or even as "acceptable" or "unacceptable") can cause significant disruptions to citizens' generally unquestioned assumption that the system is working more or less as it should. Perhaps the most significant behavior patterns will be those of governmental actors who are expected to play a role that is neutral and fair but are often accused of doing something else, such as weighting development interests more strongly than citizen concerns (Fowlkes and Miller 1987; Molotch 1976; Stone 1980). The problem may be particularly severe if the "rules of the game" appear to reflect a slant in favor of repeat players such as oil companies over one-shotters such as local residents (Galanter 1974; cf. Schnaiberg 1980). In the case of proposed OCS developments, such reactions appear to have been engendered by the widespread perception that MMS officials (particularly at higher, decision-making levels) had a decidedly pro-development bias, as well as by the reality that OCS developments constitute a multibillion-dollar source of income for the federal treasury.

Social Systems. Third, community residents may see either threats or opportunities for various aspects of community social structure that affect their lives. Potential threats include the risk of the disruptions that have often characterized large-scale industrial developments in rural areas, such as increased crime (Freudenburg 1986a; Freudenburg and Jones 1991; Krannich et al. 1984), drug and alcohol abuse (Milkman et al. 1980; Lantz and McKeown 1979), or mental health problems (Freudenburg et al. 1982; Bacigalupi and Freudenburg 1983). Threats may also be seen for characteristics of the social system that have not received the same degree of attention from researchers but are highly prized by rural residents, such as a slow-paced, peaceful, and friendly community (Dillman and Tremblay 1977), or one where "everybody knows everybody else" (Freudenburg 1986a, 1984).

As pointed out by a number of northern California coastal residents, for example, the relative scarcity of jobs in the Mendocino/Fort Bragg region meant that people really had to want to live there. The attractions included not just the coastline, but also the way of life, which was considered so valuable by locals that even real estate agents said they would not want to risk it for offshore development. As will be recalled from the interviews, the Californians expressed appreciation not just for the natural beauty of the area, but also for the sense of community that is possible in the region, in contrast with the possibilities that exist in the more urbanized areas along the more southerly sections of coastal California.

On the other hand, as we noted in the previous chapter, OCS development can also offer opportunities to maintain what residents find to be desirable social arrangements, as appears to have been the case in southern Louisiana, or to develop new forms of social relationships that appear more promising. As noted earlier, one example is provided by the concentrated work cycles, which have had the effect of permitting many of Louisiana's offshore oil workers to maintain their residences in rural areas hundreds of miles away from their place of work, while still earning high wages, an arrangement generally not feasible with traditional work scheduling (see Gramling 1989 for further discussion). Other potential opportunities have been described in the documents prepared by federal agencies and project proponents, which often tend to emphasize the potential for increased education and exposure to new ideas, for opening up ''restrictive'' social patterns, and for extending the range and level of resources available to the community (see e.g. U.S. Department of Interior et al. 1974). Comparable views are often either explicit or implicit in the opportunities that are seen by persons encouraging economic development and industrialization or community development and self-help efforts (Inkeles and Smith 1970; cf. Wilkinson et al. 1982).

Economic Systems. In contrast to the case of environmental/health systems, the implications for economic systems are often less likely to be seen in terms of threats, particularly during early discussions, than in terms of opportunities. While opportunities include potential increases in business volumes and in real estate values, primary attention is often devoted to the possibilities for new jobs. Studies do show that jobs are often created locally, although they often prove to be less significant for the local unemployment rate (Molotch 1976; Summers et al. 1976) and less attractive to local youths (Seyfrit 1986) than is commonly assumed. As noted earlier, however, when offshore oil development began in the Gulf of Mexico in the 1930s, it proceeded as a natural extension of the land-based activities. The greatest expansion occurred during what Catton and Dunlap called an ''exuberant'' era, ''when humans seemed exempt from ecological constraints'' (Catton and Dunlap 1980, 15). Given the gradual integration of offshore employment into the labor market of southern Louisiana, offshore development has continued to be defined largely as an opportunity for the intervening fifty years or more, over which time it has provided tens of thousands of coastal jobs.

Even so, OCS development can also be a source of economic threats, particularly in areas where residents are directly or indirectly dependent on environmental quality and a strong resource base. Threats can be posed for persons ranging from commercial fishermen, particularly in California and Alaska, to those whose livelihood depends on amenity-based tourism and recreation, as in Florida. Also potentially threatened are those who are living on low or fixed incomes who would have difficulties with rising costs of housing or with the higher taxes that, contrary to expectations, often accompany industrial development and growth (Real Estate Research Corporation 1974). Finally, developments can pose threats if they have the potential to "stigmatize" a region, as shown most vividly in the case of toxic or nuclear waste facilities (Slovic 1987; Kunreuther et al. 1988; Slovic et al. 1991), but as reflected also to a degree by the concerns over OCS developments among growth-oriented leaders in northern California. For a region that depends on a relatively pristine coastline as the key to its economic future, there can be economic as well as environmental threats in the potential for derricks and diesel shops.

Political/Legal Systems. Some of the most contentious of all opportunity-threat impacts come through the altercations surrounding litigation, and/or political activities, in favor of or opposed to a project or activity. The battles over OCS development off California and Florida, which have been going on for decades, provide an excellent example (National Research Council 1989). Not only have numerous lawsuits resulted from the proposed activity, but political battles have been waged at all levels, up to the highest ones. Congressional moratoria have repeatedly been used to stop lease sales off California, and eventually a presidential moratorium banned leasing off much of the east and west coast coasts until the year 2000 (Bush 1990).

The litigious nature of the debate over the opportunities and threats associated with OCS development has greatly increased the tendencies toward intractable positions in coastal communities and federal agencies. Legal processes start from the assumption that the relevant parties are adversaries, and often exacerbate the degree to which the process becomes adversarial. Once the conflict enters the arena of litigation, the lawyers for all parties insist the parties must *not* do precisely what virtually all risk communication literature insists that people *must* do, which is to talk to the persons on the other side (see, e.g., Hance et al. 1988). The net result is a worsening of what we have called in chapter 4 and elsewhere (Freudenburg and Gramling, 1993, 358) a "spiral of stereotypes." Once the information loop is severed by the litigation process, individuals on both sides begin to talk not to the other side, but about the other side. Experience shows that, in the absence of direct information, the participants will make up the information about the other side that they still want and need, so that rumors and speculation become, for both sides, the primary sources of information about presumed adversaries (cf. Coleman 1957).

Psychological Systems. Finally, particularly in extreme cases, the proposals—and the battles over the social meanings of the proposals—can create threats and/or opportunities for psychological systems, notably in terms of self-concepts and the degree to which people view themselves as effective individuals. If the powerful actors inside and outside a community seem not to care about an individual's concerns, or if parents find that, contrary to their expectations, they are unable to protect their children from developments that they find to pose unacceptable risks to health or to the environment, then impacts are created not just on the external conditions, but on some of the most sensitive, "internal" conditions of the human environment, namely, on residents' ability to see themselves as functioning participants in society. On the other hand, the residents of affected communities can come to find that not only their friends and neighbors, but they, themselves, may possess greater effectiveness than they had previously realized or thought, as in the case of one Mendocino resident who went from being the owner and manager of a bed-and-breakfast establishment to being an acknowledged regional leader who frequently calls senators and members of Congress.

One unfortunate consequence of the ways in which OCS developments have been pursued in the past is that some of the outcomes that can be defined as beneficial for at least the short-term interests of MMS (e.g., cases where opposition groups simply "give up") can also be some of the most significant in terms of their socioeconomic and psychological impacts (e.g., an increased belief that "the government doesn't care" or "people like me just don't matter"). Such a conflict in interests, however, is by no means a necessary outcome of proposals for development. Instead, it can reliably be seen as an indication of the extent to which the approaches to OCS development over the past several years have led to polarization—a set of circumstances where one "side" can only enjoy success in meeting its goals at the expense of persons on the other side—a point to which we will return in the final chapter.

Longer-Term Impacts and Overadaptation

The overwhelming majority of the social impact literature to date has involved the examination of development-stage impacts—the impacts associated with the actual development and/or operation of a project or the onset of an activity (for detailed overviews of this literature, see Freudenburg 1982, 1986b; Finsterbusch and Freudenburg, forthcoming). The same six systems of the human environment that are vulnerable to opportunity-threat impacts are relevant for the consideration of development-stage impacts. Because of the greater coverage of these impacts in existing literature, however, they will not be discussed here (see Gramling and Freudenburg 1992a for a more detailed discussion of development-phase impacts). For current purposes, we will turn instead to the longer-term impacts.

Just as past approaches have often overlooked the impacts that take place during the early or opportunity-threat phase of development, so too has there been a failure to deal systematically with longer-term issues. In general, the longer-term implications have simply been ignored, although it is possible to discern a pair of largely implicit models. The one that appears to be most common involves what could be called the "return to normal" hypothesis. The elements of this view are rarely laid out in detail, but the basic logic is that whether the short-term implications are disruptive, beneficial, or both, they are merely a temporary variation from the pre-impact conditions that will tend to reassert themselves, "naturally," once the source of disruption is removed. The other is what could called the "development" model. It reflects the belief that a period of rapid growth will be associated with "development," involving in this case the expansion of local abilities and options, including the range of persons and viewpoints as well as of economic possibilities, in what previously had been relatively "closed" communities. The longer-term expectation growing out of this perspective would be that development-induced increases in diversity would prove to be enduring ones, contributing to the communities' coping capacities even after the new industries or facilities themselves were removed from the scene.

These perspectives, however, do not provide an adequate characterization of actual "post-development" experiences—and intriguingly, there is growing evidence that they also fail to describe what actually happens in the biophysical environments that have provided the implicit models for the return-to-normal hypothesis (see e.g. Cairns 1990; Cairns and Pratt 1990). Instead, whether in communities having been hit with a Pithole-style boom-bust cycle, or in those having enjoyed a relatively long period of prosperity, what emerges after a major industry pulls out does not seem to be anything like "normalcy," either in the eyes of local residents or of outside researchers (e.g. Gulliford 1989), and neither does it reflect a continued vitality of "development."

Why not? Expectations have changed; skills have changed; earlier options are no longer available. Important leadership skills are often lost when key people leave or are transferred to more profitable locations. One community's example was noted by the Louisiana man, quoted in chapter 3, who worried that his town was "losing a top layer." As he put it, such skilled persons were precisely the kinds of "people we've been accustomed to having help us run the community," and they tend to be "hard to replace."

Even the economic flexibility that once characterized "pre-development" entrepreneurs may no longer be in existence; the old combination hardware store and bait shop, for example, may have been replaced by a specialized diesel mechanics shop. In many such cases, moreover, the owner of the new shop will not yet have paid off the loans for specialized equipment—

and the equipment may suddenly become not so much a source of prosperity as a factor in an impending bankruptcy.

What has taken place, it appears, is not actually a form of "development," at least if that term is taken to refer to an increased capability for dealing not just with a given industry, but with a wide range of challenges and opportunities. Instead, the experience provides an example of *overadaptation*. While the same six systems of the human environment that are vulnerable to opportunity-threat impacts and to development-stage impacts are also vulnerable to overadaptation, and while our technical articles have in fact discussed the longer-term impacts in terms of these six systems (see especially Gramling and Freudenburg 1992a), the following discussion will focus more narrowly on the types of impacts that have been most important for Gulf of Mexico communities, involving relationships with the biophysical environment, and elements of social and economic systems such as the acquisition of workplace skills.

The term of "overadaptation" has an important implication: Our argument is that, virtually by definition, human environments *do* "adapt" to impacts—indeed, they cannot fail to do so. The problem is not that the six systems so far identified—physical, cultural, social, economic, political, or psychological—somehow "fail" to adapt either to externally generated changes or to internally negotiated threats and opportunities. Some form of adaptation *will* take place; the relevant question has to do not so much with the possibility of adaptation but with its likely *consequences*. In most cases, the consequences are likely neither to include a return to normal or even a case of "underdevelopment," although that is what it is sometimes called in the literature from less-developed countries (cf. Bunker 1984). Instead, the potential exists for *over*specialization, or more precisely, for adapting too fully to the needs of a single extractive activity.

As biophysical ecologists have long been aware, adaptations may be either beneficial or harmful from the perspective of the adapting organism, but even adaptations that are beneficial or functional in the short run may prove to have negative consequences in the longer run, or vice versa. As first discussed many years ago, organisms in the natural world rarely seem to have evolved over time to a point of "maximum efficiency" (e.g., Bateson 1972). By definition, most of the organisms that are available for study are members of species that have managed to survive to the present. One clear implication is that there is a potential for an organism to be *too* finely tuned to a given ecological niche or given set of conditions; if not, the earth might still be the realm of the dinosaurs. The higher the efficiency and the more precise the adaptation for a given set of circumstances, unfortunately, the greater may be the organism's susceptibility to perturbations or environmental changes that bring about differing circumstances.

The same kind of potential for overadaptation can exist in the case of the human environment, as is particularly noteworthy in the case of rural communities that become heavily dependent on, or extensively adapted to, large-scale extractive industries. The very employment opportunities that are often seen as advantageous, particularly by supporters and proponents of extractive activities, can lead to the development of a dependency on a single, volatile sector of the economy, and hence create susceptibilities, in this case involving susceptibility to the vagaries of the world commodity markets for oil (see Gramling and Freudenburg 1990 or Freudenburg 1992a for more detailed discussions).

Although overadaptation can occur with almost any type of development, it appears to be much more likely to occur in extractive economies, which are frequently unable to take advantage of shared location advantages common to productive economies (support services, labor markets, etc.). Significantly, the extraction can only take place where the resources are found, and as Alaska's Prudhoe Bay oil deposits help to illustrate, the richest locations of natural resources are not necessarily the ones that would otherwise be considered the most inviting for human settlement. If extractive activities are to take place, accordingly, it is often necessary first to alter the physical, social, and economic systems of the local human environment, occasionally through projects as massive as the Trans-Alaska pipeline (cf. Gramling and Freudenburg 1992b). When the extractive activity ceases or declines, many of those alterations, even those that had been well-designed for extractive purposes, may become either obsolete or obstacles in terms of the region's new needs. To the extent to which the growth stage consumed local resources, those resources will no longer be available for new development (cf. Bunker 1984; Cronon 1992).

In essence, the problem has two aspects, and they are complementary. The first involves the loss of *other* options, and the second involves the degree of (measurable) dependence on a given industry such as oil development. The loss of other options is analogous to the potential for depletion that is often noted in debates over the extraction of natural resources. At least on a human time-scale, for example, oil is a ''nonrenewable'' resource, and even ''renewable'' resources such as fish or forests can be removed at a rate far higher than the speed at which they regenerate. What we call the ''production'' of oil, for example, actually involves taking the oil out of the ground and burning it up; by contrast, if an oil well is not ''developed,'' the oil remains in the ground, where it will still be available for future use.

The potential for depletion or loss, however, may be even greater in the case of ''resources'' that are the product of human effort—the physical capital of buildings, equipment, and so forth, and also the human and social capital of skills, knowledge, experience, teamwork, networks of supply and distribution, and so on. For most of these so-called "man-made" or anthro-pogenic resources, in contrast to the case for natural resources, the normal expectation is for depreciation, decay, or dilution over time. While it is of course possible for physical or human capital to be destroyed actively, as when a fishing dock or historic building is demolished, or when persons with unique skills or cultural characteristics are driven away from an area or even killed, the productive capacities that truly are "produced" by human effort may be the opposite of many natural resources in having the tendency to de-preciate or disappear over time *even if they are merely ignored.*

One implication is that if our analyses are expected to take into account any irreversible or irretrievable "commitment of resources," we may need to consider not only the physical loss, use or destruction of *natural* resources, but also the loss, destruction, *or even failure to maintain* the anthropogenic elements of productive capacity. Although in rare cases this danger has been recognized by local communities, and steps have been taken to protect against loss (Wybrow 1986; cf. Bowles 1981), traditional logic once held that such losses were of little concern. Traditional or pre-development activities have often been characterized officially as "underemployment," particularly in cases where people involved in traditional forms of economic activity were subsequently attracted to new activities that provided higher wages. At least with the benefit of hindsight, however, it appears that the more common prob-lem in isolated, resource-dependent communities may have been an excessive loss of pre-development capacities.

Has this been a problem in Louisiana? For decades, few local leaders would have said so, particularly given the high wages and relatively depend-able employment being provided by oil development. By the time this study's interviews were conducted in 1991, however, many leaders expressed concern about the degree to which their region had become dependent on the oil in-dustry, about the need to diversify the economy, and about the difficulty of attracting new industries to a region having a history of the high wages long associated with the oil industry. Similarly, there is a degree of independent evidence that alternative forms of economic activity have failed to be main-tained. As was noted above, Morgan City, Louisiana, the self-proclaimed "shrimp capital of the world" when offshore oil development was first be-ginning in the 1950s, no longer had a resident shrimp fleet or shrimp pro-cessing facility by the time that many residents decided they needed to "return" to shrimping as a means of maintaining their livelihood (Gramling and Brabant 1986).

The second measure of overadaptation involves the degree to which a region's economic fortunes have become tied to a single industry. In some ways, the offshore oil industry would seem particularly *un*likely to lead to overadaptation; indeed, as will be recalled, a number of Californians voiced precisely the opposite complaint, arguing that far too few of the workers would be hired locally—being brought in instead from such remote locations as coastal Louisiana.

There is reason to believe that these complaints involve more than mere boosterism. Most important is the influence of concentrated work scheduling, as noted above: Someone who works on an offshore platform for fourteen days and then has fourteen days off, for example, actually needs to "commute" to work only about once a month, thus being freed to live in almost any location having reasonable transportation facilities. One study of the offshore workers based in St. Mary parish (Gramling 1980) found that these workers actually lived in eighteen different states; for regions that might be facing their first proposals for offshore oil development today, and thus that might be expected to have less in the way of an experienced work force in place, the proportion of jobs going to nonlocal workers might be even higher. At the same time, this very combination of concentrated scheduling and widely dispersed workers might be expected to reduce the degree to which the impacts of commodity-price swings would be focused on the host region. In the case of Lafayette and St. Mary parishes, in fact, there is an important bit of evidence that the two may be less dependent on the oil industry than our discussion so far would suggest: even at the height of oil activity, less than 15% of the *direct* employment in either parish was in "mining," the standard industrial classification category that includes oil extraction (see Gramling and Freudenburg 1990).

Is overadaptation actually a significant risk for a region where more than 85% of the local labor market is apparently in other sectors? Apparently so. Given that St. Mary parish has long been a center for offshore extractive employment, while Lafayette parish has been the regional administrative center, the two parishes offer two different types of tests; more quantitative findings from both parishes, however, suggest that concerns about overadaptation should not be dismissed too hastily. While we have performed a series of tests, only the summary results will be reported here (for further discussion, see Gramling and Freudenburg 1990).

Our dependent variable is the simplest available measure of community impacts, namely total employment. Unlike *un*employment rates, which are affected by individuals' movements in and out of the labor market while not directly providing any information on such mobility, *total employment* provides a straightforward indicator of an area's economic vitality. It is also one of the most carefully measured of the publicly available statistics on economic performance. In addition, local communities frequently emphasize the importance of employment when they seek natural resource development (Summers et al. 1976; Murdock and Leistritz 1979; Molotch 1976), and within the community impact literature, employment changes tend to be seen as the driving force behind population changes and the subsequent social and economic impacts (for reviews, see Gramling 1992; Freudenburg 1986b).

To check for the possibility that findings from Lafayette and St. Mary parishes might actually reflect relatively general trends, rather than simply being limited to the energy-dependent regions of the state, we have also added a third or "control" parish. Ouachita parish, a relatively diversified parish in the north-central portion of the state, has a population comparable to that of Lafayette parish, and it was shown by an investigation of correlation coefficients to have some of the highest correlations with the substantive commodity variables of any of the parishes in the state not actually in the oil-dependent region. In non-statistical English, this means that Ouachita parish should provide the toughest double check of any comparable parish in the state.

For readers who have no taste for statistics, the next three paragraphs can be skipped; for those who want all of the details, it may be worth consulting a technical article (Gramling and Freudenburg 1990) for a fuller discussion. The next three paragraphs are intended for readers who want just the basic statistical facts and who are willing to endure a few technical terms to get them.

We have used stepwise time-series analyses to produce the results that follow. In the first step, we have used autoregressive "predictions," which make use of the common finding that the best "predictor" for a given variable (such as this year's employment in a given parish) is often the prior value of the same variable (such as last year's employment). Given that this tendency is not so much a substantive prediction as a measure of the degree of stability in the system, it is possible to compute an *index of instability*, doing so simply by comparing the *un*explained variance in total parish employment against the unexplained variance in the nation as a whole over the same time. While it is common to think of the period 1970–88 as having been a volatile one for the nation's economy, almost all of the variance in national employment for the nineteen-year period can be "explained" by the previous year's national employment (adjusted $R^2 = .9813$). The remaining, unexplained variance (.0187, or 1.87%) can be used as a benchmark; the level of variance that cannot be explained by autoregression is roughly two and a half times as high as the national average in the comparison parish of Ouachita, four and a half times as high in the headquarters parish of Lafayette, and ten times as high in St. Mary parish (unexplained variances of .0482, .0838, and .1873, respectively). This first test thus suggests that year-to-year economic instability proved to be high even in the case of the headquarters parish, but to be higher still in the case of the predominantly blue-collar parish.

In the next two steps, we added national employment figures and substantive commodity variables, with our analyses focusing on the proportional reduction of unexplained variance, or PRUV. To test for the possibility that the fluctuations in employment levels might simply reflect national economic conditions, the second step used autoregression plus the then-current year's employment of the nation as a whole. The addition of national employment trends led to a PRUV of more than 20% for Ouachita parish, but a PRUV of less than 5% for Lafayette parish, and a negative PRUV—an actual *worsening* of the adjusted R^2—for St. Mary parish. (The adjusted R^2 figures reflect the number of independent variables in the equation; the "unadjusted" R^2 increases from .790 to .792.) This clearly would not suggest that the energy-dependent parishes were simply experiencing the same conditions as the rest of the nation.

It was only in the third and final step that we added variables to represent world commodity trends. In the equation for the comparison parish, none of the commodity variables achieved statistical significance. For Lafayette and St. Mary parishes, by contrast, they were substantively as well as statistically significant: Both parishes have PRUV improvements of *more than 70 percent,* based on the commodity variables. In Lafayette, the headquarters parish, the important variables are the price of crude oil on the world market, the previous year's price, and the OPEC oil embargo of 1973–74. The autoregressive term is no longer significant, meaning that employment in Lafayette parish can actually be predicted better on the basis of the substantive, commodity-related variables than by knowing the previous year's level of employment in the same parish. The coefficient for national employment totals remains significant, suggesting that Lafayette parish did respond to a degree to the same influences that were operative for the economy as a whole, as might be expected, based on its somewhat larger and more diversified economy, but the broader economic forces and the autoregressive/moving average (ARMA) specifications, even in combination, lead to less of an improvement in explained variance once commodity variables are controlled than vice versa.

Table 7 presents the overall findings. First, the index of instability shows that the year-to-year fluctuations in employment are roughly four and a half times as bad in Lafayette parish, and ten times as bad in St. Mary parish, as in the nation as a whole over the period for which comparable data can be found (1970–88), while being only about two and a half times as bad as the national average in Ouachita, the comparison parish. Second, employment totals in Ouachita parish *do* appear to respond to national trends; those in Lafayette and St. Mary parishes do not, remaining essentially unaffected by whatever trends are evident in the rest of the country. Third and finally, substantive commodity variables—the price and consumption of oil, the world-

TABLE 7

Influences of Economic Stability, National Trends, and External/Substantive
Commodity Variables on Total Local Employment, 1970–1988

		Reductions in unexplained variance, based on addition of:		
Parish	Index of Instability[a]	National Economic Trends	Substantive Commodity Variables[b]	Variance Explained by Commodity Variables Alone[c]
Ouachita (Comparison)	258%	25.31%	−11.67%	—
Lafayette (Headquarters)	448%	4.89%	74.47%	94.01%
St. Mary (Extraction)	1002%	−14.26%	78.46%	93.76%

[a]Unexplained variance from autoregression, as ratio to comparable figures nationwide;
see text.
[b]Worldwide oil price and count of active oil-drilling rigs, U.S. oil consumption, and temporal
status vis. OPEC embargo
[c]For Lafayette parish: (734.24 × world oil price) + (3409 × previous year's price) + (3.6
× previous year's U.S. oil consumption)
For St. Mary parish: (.909 × world rig count) − (165.76 × world oil price) + (5491.38 ×
post-OPEC)

wide count of active oil-drilling rigs, and the effects of the OPEC oil embargo
of 1973–74—lead to reductions of roughly 75% in the unexplained variance
for the two oil-dependent parishes, while having no meaningful effect in Ouachita parish.

The overall conclusion is that, while direct employment in oil extraction
may not have provided more than 15% of the total employment, in either parish, for any year in this entire period, it is possible to explain more than 90%
of the *total* employment in either parish, over the roughly twenty-year period,
simply by knowing the world price of oil, the worldwide count of active oil-
drilling rigs, U.S. oil consumption, and whether the year in question came
before or after the OPEC oil embargo of 1973–74. Despite the usual exhor-
tations of economic development practitioners about the importance of "local control" over employment futures, it appears that the vast majority of the
variation in local employment was shaped by factors well outside the range of
control of such small communities—indeed, perhaps beyond the range of
control of the largest and most sophisticated of the multinational oil firms,
as well.

In short, while there is a clear need for further research, it has been possible to gain a significant number of insights based even on this simple analysis. To repeat, we have not attempted here to spell out the kinds of impacts
that are likely for all six systems of the human environment and for all three
phases of development; for such a level of detail, readers are referred to our

more technical articles (see especially Freudenburg and Gramling 1992; Freudenburg 1992a; Gramling 1992; Gramling and Freudenburg 1992a). By way to trying to offer a simple if convenient set of illustrations, however, table 8 provides a listing of some of the types of impacts that need to be considered in that future research, organized in terms of the six systems of the human environment and the three phases of development that have been described in this chapter.

TABLE 8

Longitudinal Framework for Assessing Impacts of OCS Development on the Human Environment*

EXAMPLES OF POTENTIAL IMPACTS BY PHASE OF DEVELOPMENT

System Affected	Opportunity-Threat	Development/Event	Adaptation/Post-Development
PHYSICAL	Anticipatory construction or lack of maintenance; decay of existing structures and facilities; new construction	Potentially massive alteration of the physical environment; destruction of old, and construction or upgrading of new/existing facilities	Loss of some uses due to the exploitation of others; deterioration of alternative productive facilities
CULTURAL	Initial contact; potential for loss of cultural continuity; threats to the legitimacy of existing institutions	Suspension of activities that assure cultural continuity, e.g. subsistence harvest; reduced effectiveness of traditional norms/sanctions	Gradual erosion of culture; loss of unique knowledge, skills, and/or perspectives; loss of cultural leaders, seeking jobs elsewhere
SOCIAL	Organization; investment of time, money, and energy for support or resistance; conflicts resulting from differential construction of risks	Population increases; influx of outsiders; decline in density of acquaintanceship; social change; formation of newcomer/oldtimer cleavages	Alteration of human capital, through refocus on specialized skills with few other applications; losses of organizational skills and networks
POLITICAL/LEGAL	Litigation to promote or block proposed development; intensified lobbying; organized protests; potential "civil disobedience" or even violence	Intrusion of development activity into community politics; litigation and conflict over activity impacts; decreasing capacity of community facilities and services	Recriminations over loss of earlier options and/or "unexpectedly" short duration of boom-bust prosperity; zoning/regulatory changes in search of new development
ECONOMIC	Decline or increase in property values; speculation and investment; efforts to "lock up" particularly promising parcels	Traditional boom-bust effects; inflation; entrance of "outsiders" and national chains into local labor market and retail sector	Large-scale job loss and/or unemployment; loss of economically flexible businesses; increased bankruptcies, even in "spin-off" sectors of economy
PSYCHOLOGICAL	Anxiety, stress, anger; gains or loses in perceived efficacy	Euphoria; stress associated with rapid growth; psychosocial pathology; family violence; losses or gains in efficacy	Depression and other problems associated with loss of employment; acquisition of potentially maladaptive coping strategies

*Source: Adapted from Gramling and Freudenburg 1992a.

7

Ideology and Impacts

More than a year after MMS representatives from Washington were treated to an outpouring of public criticism at Fort Bragg, a different body was holding a hearing in the same area. This body was composed of scientists who had been picked by the National Academy of Sciences, and who were later to agree with the north-coast residents rather than with the top-level personnel of MMS, concluding that the agency did not yet have enough scientific information to be able to make informed decisions concerning leasing in the area. At the time, however, this panel looked to many of the northern California residents a great deal like the others that had visited the area, and accordingly, this panel heard a set of viewpoints just as intense.

The chair of this National Academy panel was later to comment that, to him, the most succinct summary was provided by a woman, obviously bright and quite articulate, but also obviously quite frustrated, who ended her remarks with a final complaint about the leaders of MMS: "They just don't *get* it." In general, as noted above, this might be expected when differing subcultural perspectives, hardened by litigation, come into contact; but then again, perhaps the failure to "get it" has a degree of usefulness, particularly from a short-term perspective, that needs to be examined.

As we have discussed in the preceding pages, just as people in different regions can come to delineate different perspectives, so too, agencies and industries can develop distinctive cultures of their own. As a rule, the longer these cultural assumptions go without serious challenge, the more firmly entrenched they become. When serious challenges first begin, the conflicts with opposing groups can lead to further commitment to the very viewpoints that helped to create the conflicts in the first place (for a broader discussion, see Coleman 1957)—and as if the conflicts were not already polarized enough, the battles are increasingly being fought out, almost literally, in an arena that presupposes an adversarial approach to decision-making, namely the legal system. The net result is a further worsening of the spiral of stereotypes: once the information loop is severed by the threat of litigation, rumors and speculation can become the primary sources of information about the presumed adversaries in other camps (cf. Coleman 1957; see also Freudenburg and Pastor 1992).

An additional consideration, however, is that in such a polarized context, there can be a strong tendency for common beliefs to serve functions other than factuality. As we have seen repeatedly, the characterizations of the opponents to oil development that are expressed within MMS and the oil industry tend to be neither accurate nor useful—but we have been assessing their usefulness as statements of fact, rather than as ammunition in a battle. As a simple but effective rule of thumb, if people continue to cling to hypotheses that are contradicted by the available evidence—"hypotheses" that are then more commonly called by other names, such as "myths" or "ideologies"— it is often because those erroneous beliefs are useful in some other way. In a politicized context, it would scarcely be surprising if ideologies about OCS opposition continue to be embraced, with more conviction than accuracy, because in addition to being simple, plausible, and wrong, they are also politically useful.

The Political Usefulness of Selective Blind Spots?

Like most organizations, the Minerals Management Service must face a broad set of expectations—large, complex, and perhaps even inherently contradictory. It may be literally impossible for the agency to satisfy them all. As an administrator in another bureaucracy once put it, his experiences had just about convinced him that "life is a conspiracy of unfair expectations."

What do most of us do when we find it impossible to meet all of the expectations that are placed upon us? The obvious answer is that we fail to meet them all, but it may also be possible to be more specific. We may try to shift some of the responsibilities to others, as in the famous if bureaucratic rationalization, "That's not my department." We may decide to place higher priority on meeting the expectations that have the most severe consequences—where the failure to meet expectations might result in being fired, for example, rather than resulting in mere grumbling. In general, if pressed, we would probably claim that we try to deal first with the expectations we consider the most important, reasonable, or legitimate.

One of the less obvious answers, however, lies in the awareness that these very attributes—the degree to which expectations are "important," "reasonable," or "legitimate"—are not matters of scientific fact. Inherently, they involve questions of judgment, and also of political skill. One of the common characteristics of skilled bureaucratic survivors, moreover, seems to be a highly developed awareness that, if others can be convinced that an expectation is not "reasonable" or "legitimate," then the expectation can usually be avoided or ignored.

The relevance of this point to the OCS debates may already be apparent: for many years, so long as other political institutions, such as Congress and the courts, could be *successfully* convinced that people's concerns were erroneous, nonexistent, or politically and legally irrelevant, the agency was able to enjoy considerable freedom to pursue its own objectives. The technique no longer seems to work so effectively today, but perhaps that would be too much to ask from an approach that has already provided so many years of useful service.

Whatever the *accuracy* of the usual MMS and industry claims about OCS opponents—claims, for example, that the opposition is "just" a matter of misinformation, or for that matter, of irrationality, ignorance, selfishness, or a desire to commit the country to caves and candles—such assertions all tend to serve at least two additional functions. The first is that, like the other rumors about opponents that help to make up the spiral of stereotypes, they can reinforce the belief among committed partisans that they have Truth and Justice on their side. The second is that these claims also challenge the legitimacy of opponents. For purposes of political sniper fire, in fact, the central importance of such claims may have very little to do with whether or not they hit the mark, in terms of accuracy, having far more to do with whether or not they keep one's opponents pinned down, forced to fight their battles in the midst of unfavorable terrain.

Unfortunately, while the strategy of ignoring or dismissing opponents' concerns is one that can be politically efficient, at least in the short run (cf. Freudenburg and Pastor 1992), it also has its risks. In the context of OCS development, for one thing, it means that MMS has effectively been ignoring some of the very legislation under which it is supposed to operate. Importantly, the law does *not* require the agency to take only those actions that are favored in affected regions. What it does require is that if the agency's actions are likely to have negative impacts on the human environment, those impacts must be fully disclosed in advance, and not just treated as if they do not exist.

An additional problem is that, while this strategy may have simplified the short-run, day-to-day affairs of the agency's top leaders, it also tends to have worsened the social impacts that are created in affected communities and regions. Social impacts are often created, for example, when people are faced with threats over which they have little effective control, as well as when there are *conflicts* over whether a proposed development represents threats and/or opportunities. In addition, however, one of the most stressful experiences of all—particularly for red-blooded Americans who still believe the civics-textbook principle that the government is supposed to reflect the will of the governed—is to find that not just "unreasonable protesters," but they, themselves, are treated as if their most heartfelt concerns are imaginary or irrelevant (Krauss 1989; Brown and Mikkelsen 1990; see also Levine 1982). In a number of such cases, moreover, people *do* react with increasing frustration, even rage, especially if they are repeatedly ignored or treated with condescension or contempt. The ironic net result can be that the agency's opponents truly can start to sound and act "emotional," but that they can do so as a direct result of the agency's own actions.

It may in fact be true that some proposed developments genuinely do reflect the public interest so well that they can only be opposed by those who are ignorant or selfish. Proponents may also believe that the same characteristics exist in other cases where the facts are more ambiguous. This approach to the "framing" of debates has such an obvious potential for strategic usefulness, however, that at a minimum, its accuracy needs to be treated as a question for analysis, not a matter of "fact."

In reality, of course, both the views within the agency and those within the affected region are likely to have at least *some* basis in fact. Few if any changes are ever likely to be "all good" or "all bad," but in a polarized debate, few if any of the active participants seem inclined to admit to the weaknesses of their arguments unless absolutely necessary. Conflicts and stresses are thus likely to persist despite (or in some cases partly because of) the fact that proponents will insist that the implications of development are virtually all good, while opponents will counter, with apparently equal conviction, that the implications are nearly all bad.

By their very nature, moreover, negotiations over meanings tend not to offer circumstances where it is possible to "prove" one side or another to be true. As a simple if effective rule of thumb, technological controversies may be most likely to arise in cases where at least a certain degree of ambiguity exists; if all parties agree on what "the facts" are, conflict is much less likely to endure or grow. Yet the issue of "proof" tends to be muddied by at least three additional factors.

The first of those factors is that few mortals, whether they be normal citizens or even scientific geniuses, are capable of keeping in mind *all* of the facts that are reasonably relevant to policy decisions about complex technologies. Even in cases where the relevant considerations could all be known and remembered, the existence of a controversy would suggest that differing groups were strongly divided in terms of which subsets of the facts they see as most relevant. It follows, accordingly that there can be great power in the ability to influence *which* facts are given the greatest weight in policy debates—a point that is reinforced by the two remaining factors.

The second factor is that, if any side in a given debate can succeed in having its version of the facts *accepted* as accurate, that side can normally prevail. It is a virtual truism of politics and propaganda that if an assertion or a framing of an issue is repeated often enough, without being countered by alternative views, then that assertion or framing can come to be accepted or taken for granted. Given that ignorance of such truisms is rarely taken to be an asset for top-level agency officials, it would scarcely be surprising if those officials were tempted to be especially persistent in repeating, in a matter-of-fact way, the very perspectives and framings that might be most useful for their intended goals.

The third of the complicating considerations is that, even if all other factors could be equal, up to the point that two or more views of the opportunities and threats would be equally accurate, pronouncements from "official" bodies such as government agencies will normally enjoy the benefit of a doubt, while those of their critics generally will not. Such an arrangement may not cause significant quandaries in cases where disagreements truly do come down to a black-and-white, true-and-false judgment, so long as citizens and the media are able to discern "the real truth." It can create significantly greater distortions, however, when even "the facts" involve considerable ambiguity, when the relevant facts are so complex and numerous that it may be virtually impossible for most mortals to keep them all in mind at once, and when, as much as we may pretend to the contrary, the decisions inherently depend as much on values and frames of reference as they do on the facts themselves.

In the case of offshore development, as with so many other controversies over technological risk, the greatest simplification *possible* is to reduce at least the risk-related issues to two questions, not one. The first, which at least in principle could be answered scientifically, is "How safe would it be?" The second, which *cannot* be answered except as a matter of values and individual judgments, is "Is that safe enough?" (Freudenburg 1988, 1993; cf. Rayner and Cantor 1987). There is also a third question, although it is most often left unasked: "What are the blind spots that result from our current frame of reference—or from spin doctors' efforts at diversionary reframing?"

Perhaps the most common strategy among the proponents of offshore development, particularly up to the recent past, has been to assert that all relevant questions, in effect, "have been answered scientifically," but the assertion no longer has the credibility that it might once have enjoyed. The National Academy of Sciences, as noted above, has come to a different conclusion about the adequacy of the agency's scientific research on the facts involved, and as just pointed out, a "scientific" answer to questions about values and framings would be a contradiction in terms. A more realistic view might be that, to date, only a subset of the potentially relevant factual questions have indeed been answered, and that many of the key questions about values and blind spots are still awaiting a resolution.

Conclusion: Oil on the Edge

The perceived risks of offshore oil and gas development, while intensely salient both within the MMS and within the affected regions, may simply be the most visible of the impacts that OCS-related activities have already had upon the human environment. These and other opportunity-threat impacts have al-

ready had real and tangible effects, up to and including the formation of social movements, the emergence of large-scale protests and demonstrations,
and the imposition of congressional and presidential moratoria. While these
impacts have often taken place even in regions where little or no physical dis-
turbance is yet apparent, they are no less "real" for that reason—and as the
agency's growing difficulties with public reactions seem to indicate, they are
no more safely ignored.

A number of the concepts employed in this book, such as the
opportunity-threat impacts of OCS development, have received little attention in the OCS debates to date. In part, this may reflect the fact that they
have a certain resemblance to physical-science concepts such as gravity: they
cannot be directly measured, even though they do tend to have measurable
consequences. This means, as MMS has increasingly been learning, that such
impacts can create significant difficulties for those who take the risk of trying
to wish them away. It is no more true in dealing with social impacts than in
standing in front of a speeding oil truck that "what you don't see can't hurt
you." Even "mere perceptions," to reiterate one of the best-known observations in the social sciences, tend to be very real in their consequences.

A point that bears repeating here, however, is that the risks of OCS development lead not just to "expected" impacts, but to *actual* impacts. They
involve significant, empirically verifiable changes that would not have taken
place but for the announcement of a proposed development, the actions that
were taken to encourage, discourage, or otherwise influence the outcome of
the proposed development, and the social definitions of the development that
emerge as a result of the unfolding negotiation processes. In these negotiations, moreover, project proponents and government officials are far from being impartial observers; in fact they are key participants.

That seems to be part of the problem. In an earlier era, or in an
extraction-dependent region such as Louisiana, where OCS developments
created few conflicts with established uses, it might have been true that concerns about offshore developments would have been limited to a small, unrepresentative minority. It might even have been true, at least for short-term
political purposes, that the "efficient" way to deal with such concerns would
have been to dismiss them rather than to take them seriously. Today, however,
as the interviews indicate, to make such assertions is to provide evidence, to
a broad cross-section of the citizenry, that the responsible parties are acting
irresponsibly.

Whether the focus is on Louisiana or California, there appear to be no
"silent majorities," hidden from sight. The people in Louisiana truly *are* as
supportive of oil development as they seem, and those in northern California
are just as strongly opposed. Whether the focus is on the historic patterns of
support for oil development in coastal Louisiana, the vehement opposition in

California, or both, the local views are quite sensible reflections of the sociohistorical realities of the people and places in question.

Offshore oil development was accepted quite readily, even eagerly, by Louisiana coastal residents during the first half of the twentieth century, but barring changes just as dramatic as those that have taken place between the Louisiana of 1929 and the California of 1992, there is very little likelihood that the early experiences will be repeated in California, or for that matter in most other coastal regions of the United States, as we end the twentieth century and enter the twenty-first. Louisiana's rapid and ready acceptance of offshore oil reflected not any universal appeal of oil development, but a specific combination of historical, biophysical, and sociocultural factors—a set of circumstances that would be hard to replicate in virtually any of the coastal regions of the United States today, and certainly in northern California.

This reality has often been misunderstood by environmentalists, while being staunchly resisted by spin-control specialists within MMS and the oil industry. Given that some of the relevant parties are extremely skilled, despite the frequent lack of success of their "public education" efforts to date, it should perhaps not be a surprise to learn that the spin-control specialists have done more than just to emphasize the opportunities, rather than the threats, of oil development. Instead, the proponents have also tried to dismiss (or in the jargon of the field, to "de-legitimate"), rather than to deal with, the intrinsically legitimate concerns that have been expressed by a broad spectrum of the citizenry.

The results, however, have been rich in irony. Industry and agency spokespersons have repeatedly claimed to have facts and rationality on their side, but at least when they have been characterizing their opponents, these spokespersons have been far more insistent than accurate. At the same time, while their goals presumably have included an increased rate of leasing off the California coast, they seem to have succeeded not so much in overcoming the opposition as in solidifying it.

Perhaps these proponents have indeed seen their projects as genuinely offering just opportunities, and not threats. Perhaps they genuinely do believe that their opponents are irrational or misinformed. Whether such a "spin" is genuine or strategic, however, it has implications that cannot safely be ignored. While the strategy is often politically successful—although it appears to have enjoyed spectacularly little success in northern California—it is also one that can be socially stressful, worsening some of the key social impacts of development. When MMS and oil-industry representatives characterize opponents to offshore development as being selfish, ill-informed, or irrational, these characterizations themselves have impacts. Although often not recognized as such in the past, these characterizations actually constitute a key part of the process by which meanings are socially negotiated and constructed; often, unfortunately, they serve to exacerbate the opportunity-threat impacts that are created for affected communities.

To the degree to which evidence is becoming available from empirical studies—not just related to offshore oil development, but also related to technological controversies more broadly—it suggests that the views of affected citizens tend to be more rational than such characterizations would claim (Mitchell 1984; Dunlap and Olsen 1984; Szasz 1990; Slovic et al. 1984; Freudenburg 1993). At the same time, the official positions—including the most carefully developed assessments of purportedly "real" risks—are being found to be subject to far greater errors than previously suspected (Perrow 1984; Clarke 1988b; Freudenburg 1988, 1992b). While it would be premature to conclude that the general pattern definitely does hold in the case of OCS development, there is growing evidence that assessments of the risks of offshore development have been flawed by some of the same failures to consider human and organizational factors that have been found to lead to errors in other contexts (Clarke 1990; Paté-Cornell 1990; Paté-Cornell and Bea 1992; see also Royal Commission on the Ocean Ranger Marine Disaster 1984). In short, past thinking about these problems has been sufficiently deficient as to worsen them, rather than to help move the debate toward resolution.

The notion that government agencies respond to "real" risks and opportunities, while citizens are reacting mainly to (implicitly erroneous) "perceptions," may be popular in the subcultures of the agencies in question, but it is simply one that cannot be supported in the real world. Far from playing the neutral or mediating role they often attempt to project, government agencies are frequently key actors and even proponents. Agencies such as MMS often initiate the process (Broadbent 1989; Kunreuther et al. 1982), do their best to determine which issues and topics will be considered legitimate (Wynne 1982), and largely shape as well the processes by which other social actors such as the media will decide what issues to discuss and how (Freudenburg et al. 1991; Molotch 1970; Stallings 1990). In fact, to underscore a point that by now should be clear, perceptions of risks are *socially* constructed, *both* in the agency subcultures *and* in the affected communities.

Yet the important failures are not simply those of the agency and the industry. In the case of the agency, the temptation to avoid the messy questions of OCS-related social impacts, while understandable, does appear to have been a bit too strong; the law, after all, is scarcely ambiguous in its requirement for MMS to assess, and not to attempt to dismiss, its impacts on the human environment. In the case of the industry, it should probably be *expected* that the proponents of development would want to make the strongest case they can; perhaps the fairest criticism, accordingly, is that they have been carrying out their tasks with a bit more enthusiasm than accuracy. Unlike the muckrakers of old, we have seen no overt evidence of cases where oil-industry interests have been blackmailing or bribing the officials or the media; they have been playing the well-known American game of politics, and they have been following the well-worn (if not always accurate) precept that the best defense is a good offense. If their environmental opponents have been less successful in offering reframings of their own—or if those of us who have been studying the fray have been slow to recognize blind spots of our own—that is scarcely the fault of the oil industry or of MMS.

To reiterate another important point, accordingly, at least part of the blame belongs to those of us who have attempted to analyze the policy problems to date. Like our earlier points, this one should not be overblown, particularly given that the present book grows out of what may literally be the first social science study, outside of those dealing with economics and planning, to be funded by MMS in the lower forty-eight states; it is thoroughly understandable, under the circumstances, that the body of available concepts, as well as of findings, would be small. The key is to realize that, to the degree to which our goal is an improved and more balanced understanding of the ongoing debates—as opposed to "success" in promoting or opposing a given development project—we need to do better. We need to *think* better.

This is a conclusion that extends well beyond the actions of the Minerals Management Service. In all too many cases, to date, agencies and project proponents have taken advantage of the ambiguity of past terminology, insisting that they have seen no need to deal with impacts that are "merely perceptual," being "anticipatory" rather than real, or (purportedly) being so far in the future as to be "beyond our control."

In empirical fact, as is becoming increasingly clear, these impacts have often proved to be every bit as real, as quantifiable, as predictable, and as significant, as the development-phase impacts that have been officially acknowledged. Given that impacts do not cease to exist if they are simply ignored, the failure to deal with the broader range of impacts has effectively meant that, rather than *dealing* with risks, we have simply transferred them, shifting them from the principal beneficiaries of development "to local communities and residents who are little more than innocent bystanders" (Freudenburg and Gramling 1992, 952).

Too often in the past, in sum, the fuller range of social impacts have been not so much "beyond our *control*" as "beyond our *concepts*." In the case of many projects, it has been only one small step to treating these real and predictable impacts as having been "beyond our responsibilities" as well.

Such selective blind spots can offer considerable convenience for some, but they do so at the expense of avoidable, adverse impacts for others—and at the expense of accuracy. The logical reason for assessing the impacts of development, whether on the human, coastal, or marine environment, goes beyond compliance with the law; it is to assist in the making of better-informed decisions, doing so in part by providing better information. If the promise of impact assessment is to be fulfilled, however, that information needs to deal with the full range of impacts that are significant in their consequences—not just those that are conceptually convenient.

As we move into the twenty-first century, having just been involved in an oil-induced war in the Persian Gulf, we need better tools for the decisions ahead. For over four decades, offshore development in the United States has been treated primarily as a short-term economic and a political issue. Even within these restricted arenas, situations have frequently been misread. Offshore development involves complex environmental, cultural, social, economic, political, and even psychological issues. If we are to extricate ourselves from the current impasse, it will require solutions that emerge out of the balanced consideration of the inputs and concerns of all who are affected, not just from the partisans of a particular point of view.

APPENDIX: THE LEGAL CONTEXT

The practical importance of public reactions has become virtually self-evident to an increasing number of observers of the OCS policy scene, but the laws of the United States provide additional reasons for the MMS, in particular, to devote additional attention to the topic. Two laws are particularly relevant; the first of them is the National Environmental Policy Act of 1969, or NEPA (P.L. 91–190, 42 U.S.C. 4321 et seq.)[1] One of the central requirements of NEPA is that before any agency of the federal government may take "actions significantly affecting the quality of the human environment," the agency must first prepare an Environmental Impact Statement, or EIS. Section 101 of NEPA notes that the purpose of the act is not only to maintain environmental quality, but also to "fulfill the social, economic and other requirements" of U.S. Citizens. Section 102(2)(A) of the act requires federal agencies to make "integrated use of the natural *and social* sciences . . . in decision-making which may have an impact on man's environment" (emphasis added).

The President's Council on Environmental Quality (CEQ), which was given responsibility for overseeing the implementation of the law, has provided further indications that social science input is needed. Section 1508.8 of the Regulations for Implementing NEPA (U.S. Council on Environmental Quality 1978) notes that EISs are to consider direct and indirect social and cultural impacts, as well as environmental impacts. Section 1508.14 of the regulations notes that, while social and economic effects by themselves do not require preparation of an EIS, "When an environmental impact statement *is* prepared" because of impacts on the biophysical environment, and when the social and the other environmental impacts are interrelated, "then the environmental impact statement will discuss *all* of these effects upon the human environment" (U.S. Council on Environmental Quality 1978, 29, emphasis added; see also Atherton 1977; Catalano et al. 1975; Freudenburg and Keating 1985; Jordan 1984; Meidinger and Freudenburg 1983; Pring 1981; Savatsky 1974).

As the reference to "integrated use" of the social sciences makes clear, NEPA requires that agencies consider all of the significant species involved in an ecosystem, including *Homo sapiens*. The CEQ Regulations for Imple-

147

menting the Procedural Provisions of the National Environmental Policy Act (40 C.F.R. 1508.14) make the point clearer still. The "human environment" is to be "interpreted comprehensively," to include "the natural and physical environment and the relationship of people with that environment." Agencies need to assess not just so-called "direct" impacts or effects, but also "aesthetic, historic, cultural, economic, social, or health" impacts, "whether direct, indirect, or cumulative" (40 C.F.R. 1508.8).

Contrary to what many people believe, NEPA does *not* require an agency to avoid any actions that might have negative impacts, and neither does it require any kind of explicit risk–benefit trade-off. What it does require is that the agency take a "hard look" at potentially negative implications, making "integrated use" of the social as well as biophysical sciences, and exploring options for mitigating any negative impacts that are identified. The EISs thus are intended to provide a kind of full disclosure for federal decision-makers, who are then expected to consider the negative as well as the positive implications of potential courses of action before they proceed. NEPA also provides citizens with the opportunity to challenge agency decisions in court; again in this case, however, NEPA's provisions are often misunderstood; the greatest level of legal vulnerability for the agency is created *not* by taking actions that have negative impacts, but by failing to give full, good-faith consideration to those negative impacts in the relevant environmental impact statement.

Court decisions involving social impacts have followed this interpretation. On the one hand, decisions all the way up to the Supreme Court (most notably in *Metropolitan Edison v. People Against Nuclear Energy,* 103 S. Ct. 1556 [1983]) have consistently held that social and economic impacts, *by themselves,* do not require the preparation of EISs (see also the discussion in Llewellyn and Freudenburg 1990). Like the CEQ regulations, court decisions have generally drawn no distinction between the "social" and the "economic" halves of the impacts on the human environment—or "socioeconomic" impacts, as they are commonly called—but a number of court decisions have been particularly clear in noting that *economic* impacts, in themselves, are insufficient to require the preparation of EISs (for further discussion, see Llewellyn and Freudenburg 1990). In short, an EIS needs to be prepared only if an agency's actions will have significant physical and/or biological consequences.

Once an EIS *does* need to be prepared, however, agencies such as MMS are required to assess *all* of the significant impacts on the environment, including those that affect human beings. One of the better-known rulings to that effect, again in accordance with the CEQ regulations, actually involved the Department of Interior: the department had prepared an EIS as part of a major coal sale in the Powder River Basin of Wyoming and Montana; the EIS was challenged by the Northern Cheyenne tribe for failing to discuss the likely social, economic, and cultural impacts on the tribe. The court's strongly worded decision—*Northern Cheyenne Tribe v. Hodel*, no. CV 82-116-BLG (D. Mont. May 28, 1985)—overturned the EIS, chastised the Department of Interior for its failure to convert an "ostensible concern" with social impacts into "any meaningful analysis of the extent of such impacts" (p. 14), and voided the sale of over 350 million tons of coal with a market value of well over four billion dollars. The department chose not to appeal, and instead commissioned the studies that had not previously been done, ultimately issuing a supplemental EIS.

A second law, the Outer Continental Shelf Lands Act (43 U.S.C. 1331 et seq.), as modified by the Outer Continental Shelf Lands Act Amendments, or OCSLAA (P.L. 95-372, 43 U.S.C. 1801 et seq.), makes it even clearer that assessments of OCS activities, in particular, require explicit attention to social and economic impacts. The law specifically states that "The Secretary [of Interior] shall conduct a study of any area or region included in any oil and gas lease sale in order to establish the information needed for assessment and management of environmental impacts on the human, marine, and coastal environments of the outer continental shelf and the coastal areas which may be affected by oil and gas development in such area or region" (43 U.S.C. 1346 (a)(1)). The "human environment" is given even a stronger definition in the statutory language of OCSLAA itself than in the CEQ regulations on NEPA: "The term 'human environment' means the physical, social, and economic components, conditions and factors which interactively determine the state, condition, and quality of living conditions, employment, and health of those affected, directly or indirectly, by activities occurring on the outer continental shelf" (43 U.S.C. 1331 (i)).

Finally, the OCSLAA provisions also require the Secretary of Interior to "conduct such additional studies to establish environmental information as he deems necessary and shall monitor the human, marine, and coastal environments of such area or region in a manner designed to provide time-series and data trend information which can be used for comparison with any previously collected data for the purpose of identifying any significant changes in the quality and productivity of such environments, for establishing trends in the areas studied and monitored, and for designing experiments to identify the causes of such changes" (43 USC 1336 20 (b)). Again, this is an even stronger and more explicit set of requirements than those that are imposed on all federal activities under NEPA. In general, however, the research arm of the National Academy of Sciences has concluded that the law's provisions for attention to "the human environment" have not been fulfilled by Interior (National Research Council 1989, 1992, 1993).

REFERENCES

Atherton, Carol Coop. 1977. "Legal Requirements for Environmental Impact Report-
ing." In James McEvoy III and Thomas Dietz, eds. *Handbook for Environmental
Planning: The Social Consequences of Environmental Change*, 9–64. New York:
Wiley.

Bacigalupi, Linda M. and William R. Freudenburg. 1983. "Increased Mental Health
Caseloads in an Energy Boomtown." *Administration in Mental Health* 10 (4,
Summer): 306–22.

Bateson, Gregory. 1972. *Steps to an Ecology of Mind*. New York: Ballantine.

Bernstein, Marver H. 1955. *Regulating Business by Independent Commission*. Princeton, NJ: Princeton University Press.

Blair, John M. 1976. *The Control of Oil*. New York: Vintage.

Block, Fred. 1987. *Revising State Theory: Essays in Politics and Postindustrialism*.
Philadelphia: Temple University Press.

Botzum, John R. and Diane K. Garner. 1988. "Why Can't the Interior Dept. Get Its
Story Across?" *Coastal Zone Management* 19 (4, Feb. 10): 1–3.

Bowles, Roy T. 1981. "Preserving the Contribution of Traditional Local Economies."
Human Services in the Rural Environment 6(1):16–21.

Brantly, J. E. 1971. *History of Oil Well Drilling*. Houston: Gulf Publishing Co.

Broadbent, Jeffrey. 1989. "Strategies and Structural Contradictions: Growth Coali-
tion Politics in Japan." *American Sociological Review* 54 (October): 707–21.

Brown, Phil and Edwin J. Mikkelsen. 1990. *No Safe Place: Toxic Waste, Leukemia,
and Community Action*. Berkeley, CA: University of California Press.

Bunker, Stephen G. 1984. "Modes of Extraction, Unequal Exchange, and the Progressive Underdevelopment of an Extreme Periphery: The Brazilian Amazon,
1600–1980." *American Journal of Sociology* 89 (March):1017–64.

Bush, George H. 1990. Statement by the President [Press release related to decision
to extend moratoria on OCS activities in Florida, California, and a number of
other contentious sites], Washington, D.C., June 26.

Cairns, John, Jr. 1990. "Lack of Theoretical Basis for Predicting Rate and Pathways of Recovery." *Environmental Management* 14 (5): 517–26.

Cairns, John, Jr. and James R. Pratt. 1990. "Biotic Impoverishment: Effects of Anthropogenic Stress." In George M. Woodwell, ed., *The Earth in Transition: Effects of Anthropogenic Stress*,495–505. Cambridge: Cambridge University Press.

Caldwell, Lynton K. 1982. *Science and the National Environmental Policy Act: Redirecting Policy through Procedural Reform.* University, AL: Univ. Alabama Press.

Catalano, Ralph, Stephen J. Simmons, and Daniel Stokols. 1975. "Adding Social Science Knowledge to Environmental Decision Making." *Natural Resources Lawyer* 8 (1): 41–59.

Catton, William R., Jr. 1989. "Cargoism and Technology and the Relationship of these Concepts to Important Issues such as Toxic Waste Disposal Siting." In Dennis L. Peck, ed., *Psychosocial Effects of Hazardous Toxic Waste Disposal on Communities*, 99–117. Springfield, IL: Charles Thomas.

Catton, William R., Jr. and Riley E. Dunlap. 1980. "A New Ecological Paradigm for Post-Exuberant Sociology." *American Behavioral Scientist* 24 (Sept./Oct.): 15–47.

Child, J. 1981. "Culture, Contingency, and Capitalism in the Cross National Study of Organizations." In B. Staw and L. L. Cummings, eds., *Research in Organizational Behavior,* 303–57. Greenwich, CT: JAI.

Cincin-Sain, Biliana and Robert Knecht. 1987 "Federalism Under Stress: The Case of Offshore Oil and California." In Harry Scheiber, ed., *Perspectives on Federalism: Papers From the First Berkeley Seminar on Federalism*, 149–76. Berkeley: University of California.

Clark, Burton R. 1968. *Adult Education in Transition: A Study of Institutional Insecurity.* Berkeley: University of California Press.

Clark, John G. 1987. *Energy and the Federal Government: Fossil Fuel Policies, 1900–1946.* Urbana: University of Illinois Press.

Clarke, Lee. 1988a. "Explaining Choices Among Technological Risks." *Social Problems* 35 (1, February): 22–35.

————. 1988b. "Politics and Bias in Risk Assessment." *The Social Science Journal* 25 (2): 155–65.

————. 1990. "Organizational Foresight and the Exxon Oil Spill." Paper presented to annual meeting of Society for Study of Social Problems, Washington, D.C., August.

Cole, R. E. 1979. *Work, Mobility, and Participation: A Comparative Study of American and Japanese Industry.* Berkeley: University of California Press.

Coleman, James S. 1957. *Community Conflict*. Glencoe, IL: Free Press.

Coleman, James W. 1989. *The Criminal Elite: The Sociology of White Collar Crime* (Second Edition). New York: St Martin's Press.

Comeaux, Malcolm L. 1972. *Atchafalaya Swamp Life: Settlement and Folk Occupations*. Baton Rouge: Louisiana State University Press.

Conservation Foundation, The. 1987. *Groundwater Protection*. Washington, D.C.: The Conservation Foundation.

Creighton, James L. 1980. *Public Involvement Manual: Involving the Public in Water Power Resources Decisions*. Washington, D.C.: U.S. Government Printing Office.

Cronon, William. 1992. "Kennecott Journey: The Paths Out of Town." In William Cronon, George Miles and Jay Gitlin, eds., *Under an Open Sky: Rethinking America's Western Past*, 28–51. New York: Norton.

Darrah, William C. 1972. *Pithole, The Vanished City*. Gettysburg, PA: Published by the Author.

Deal, T. E. and A. A. Kennedy. 1982. *Corporate Cultures*. Reading, MA: Addison-Wesley.

Dillman, Don A. and Kenneth R. Tremblay, Jr. 1977. "The Quality of Life in Rural America." *Annals, AAPSS* 429 (January): 115–29.

Dunlap, Riley E. 1987. "Polls, Pollution, and Politics Revisited: Public Opinion on the Environment in the Reagan Era." *Environment* 29 (July/August): 6–11, 32–37.

———. 1992. "Trends in Public Opinion Toward Environmental Issues: 1965–1990." In Riley E. Dunlap and Angela G. Mertig, eds., *American Environmentalism: The U.S. Environmental Movement, 1970–1990*, 89–116. New York: Taylor and Francis.

Dunlap, Riley E. and William R. Catton, Jr. 1979. "Environmental Sociology." *Annual Review of Sociology* 5:243–73.

Dunlap, Riley E. and Marvin E. Olsen. 1984. "Hard-Path versus Soft-Path Advocates: A Study of Energy Activists." *Policy Studies Journal* 13:413–28.

Durio, C. and K. Dupuis. 1980. "Public Utilities." In R. Gramling, ed., *East St. Mary Parish, Economic Growth and Stabilization Strategies*, 258–77. Baton Rouge: Louisiana Department of Natural Resources.

Edelstein, Michael R. 1988. *Contaminated Communities: The Social and Psychological Impacts of Residential Toxic Exposure*. Boulder, CO: Westview.

Edmunds, James, ed. 1983. *The Times of Acadiana*. Lafayette, LA.

Engler, Robert. 1961. *The Brotherhood of Oil: Energy Policy and the Public Interest.* Chicago: University of Chicago Press.

Erikson, Kai T. 1976. *Everything in Its Path: The Destruction of Community in the Buffalo Creek Flood.* New York: Simon and Schuster.

Feagin, Joe R. 1985. "The Global Context of Metropolitan Growth." *American Journal of Sociology* 90:1204–30.

————. 1990. "Extractive Regions in Developed Countries: A Comparative Analysis of the Oil Capitals, Houston and Aberdeen." *Urban Affairs Quarterly* 25 (4, June): 591–619.

Festinger, Leon. 1957. *A Theory of Cognitive Dissonance.* Evanston, IL: Row, Peterson.

Finn, Kathy. 1992. "Where Are They Now? New Orleans Area S&Ls." *New Orleans Magazine* 26 (4, Jan.): 38–39.

Finsterbusch, Kurt. 1988. "Citizens' Encounters with Unresponsive Authorities in Obtaining Protection from Hazardous Wastes." Presented at the Annual Meeting of the Society for the Study of Social Problems, Atlanta, August.

————. 1989. "Community Responses to Exposures to Hazardous Wastes." In Dennis L. Peck, ed., *Psychosocial Effects of Hazardous Toxic Waste Disposal on Communities,* 57–80. Springfield, IL: Charles C. Thomas.

Finsterbusch, Kurt F. and William R. Freudenburg. Forthcoming. "Social Impact Assessment and Technology Assessment." Riley Dunlap and William Michelson, eds., *Handbook of Environmental Sociology.* Westport, CT: Greenwood.

Fowlkes, Martha R. and Patricia Y. Miller. 1987. "Chemicals and Community at Love Canal." In Branden B. Johnson and Vincent T. Covello, eds., *The Social and Cultural Construction of Risk: Essays on Risk Selection and Perception,* 55–78. Dordrecht, Holland: D. Reidel.

Freeman, A. Myrick, III and Robert H. Haveman. 1972. "Clean Rhetoric and Dirty Water." *The Public Interest* 28 (Summer): 51–65.

Freitag, Peter J. 1975. "The Cabinet and Big Business: A Study of Interlocks." *Social Problems* 23:137–152.

————. 1983. "The Myth of Corporate Capture: Regulatory Commissions in the United States." *Social Problems* 30:480–91.

Freudenburg, William R. 1982. "The Impacts of Rapid Growth on the Social and Personal Well-Being of Local Community Residents," In Bruce A. Weber and Robert E. Howell, eds., *Coping with Rapid Growth in Rural Communities,* 137–70. Boulder, Colorado: Westview Press.

———. 1984. "Boomtown's Youth: The Differential Impacts of Rapid Community Growth Upon Adolescents and Adults." *American Sociological Review* 49 (5, October): 697–705.

———. 1986a. "The Density of Acquaintanceship: An Overlooked Variable in Community Research?" *American Journal of Sociology* 92 (1, July): 27–63.

———. 1986b. "Social Impact Assessment." *Annual Review of Sociology* 12: 451–78.

———. 1988. "Perceived Risk, Real Risk: Social Science and the Art of Probabilistic Risk Assessment." *Science* 242 (October 7): 44–49.

———. 1991. "Rural-Urban Differences in Environmental Concern: A Closer Look." *Sociological Inquiry* 61 (2, May).

———. 1992a. "Addictive Economies: Extractive Industries and Vulnerable Localities in a Changing World Economy." *Rural Sociology* 57 (3, Fall): 305–32.

———. 1992b. "Nothing Recedes Like Success? Risk Analysis and the Organizational Amplification of Risks." *Risk* 3 (#1, Winter): 1–35.

———. 1993. "Risk and Recreancy: Weber, the Division of Labor, and the Rationality of Risk Perceptions." *Social Forces.* 71(4, June): 909–32.

Freudenburg, William R., Linda M. Bacigalupi, and Cheryl Landoll-Young. 1982. "Mental Health Consequences of Rapid Community Growth: A Report from the Longitudinal Study of Boomtown Mental Health Impacts." *Journal of Health and Human Resources Administration* 4 (3, Winter): 334–52.

Freudenburg, William R., Cynthia-Lou Coleman, Catherine Helgeland, and James Gonzales. 1991. "Media Coverage of Hazard Events." Presented at Annual Meeting, Society for Risk Analysis.

Freudenburg, William R. and Robert Gramling. 1990. "Community Impacts of Technological Change: Toward a Longitudinal Perspective." Presented at Annual Meeting of Rural Sociological Society, Norfolk, VA, August.

———. 1992. "Community Impacts of Technological Change: Toward a Longitudinal Perspective." *Social Forces* 70 (4): 937–55.

———. 1993. "Socio-Environmental Factors in Resource Policy: Understanding Opposition and Support for Offshore Oil Development." *Sociological Forum* 8 (#3, Sept.): 341–64.

———. 1994. "Bureaucratic Slippage and Failures of Agency Vigilance: The Case of the Environmental Studies Program." *Social Problems* 41 (#2, May).

Freudenburg, William R. and Timothy R. Jones. 1991. "Attitudes and Stress in the Presence of Technological Risk: A Test of the Supreme Court Hypothesis." *Social Forces* 69 (4, June): 1143–68.

Freudenburg, William R. and Kenneth M. Keating. 1985. "Applying Sociology to Policy: Social Science and the Environmental Impact Statement." *Rural Sociology* 50 (4): 578–605.

Freudenburg, William R. and Susan K. Pastor. 1992. "Public Responses to Technological Risks: Toward a Sociological Perspective." *Sociological Quarterly* 33 (3, August): 389–412.

Friesema, H. Paul and Paul J. Culhane. 1976. "Social Impacts, Politics, and the Environmental Impact Statement Process." *Natural Resources Journal* 16:339–56.

Frost, P. J., L. F. Moore, M. R. Louis, C. C. Lundburg, and J. Martin. 1985. *Organizational Culture.* Beverly Hills: Sage.

Galanter, Marc. 1974. "Why the 'Haves' Come Out Ahead: Speculations on the Limits of Legal Change." *Law and Society Review* 9:95–160.

Gallaway, Bennie J. 1984. "Assessment of Platform Effects on Snapper Populations and Fisheries," Pp. 130–137 in *Proceedings of the 5th Annual Information Transfer Meeting.* New Orleans: U.S. Department of Interior/Minerals Management Servide (MMS 85-0008).

Gamson, William A. and Andre Modigliani. 1989. "Media Discourse and Public Opinion on Nuclear Power: A Constructionist Approach." *American Journal of Sociology* 95 (1, July): 1–37.

Ghanem, Skukri. 1986. *OPEC: The Rise and Fall of an Exclusive Club.* London: Routledge and Kegan Paul.

Glasser, Barney G. and Anselm L. Strauss. 1967. *The Discovery of Grounded Theory: Strategies for Qualitative Research.* Chicago: Aldine.

Gould, Gregory J. 1989. *OCS National Compendium.* Washington, DC: U.S. Minerals Management Service.

Gould, Gregory J., Robert M. Karpas, and Douglas L. Slitor. 1991. *OCS National Compendium.* Herndon, VA: U.S. Minerals Management Service.

Gramling, Robert. 1980. "The Economic History of East St. Mary Parish." In R. Gramling ed., *East St. Mary Parish, Economic Growth and Stabilization Strategies,* 4–16. Baton Rouge: Louisiana Department of Natural Resources.

————. 1983. "A Social History of Lafayette Parish." In David P. Manuel, ed., *Energy and Economic Growth in Lafayette, LA: 1965–1980,* 8–52. Lafayette, LA: The University of Southwestern Louisiana.

————. 1989. "Concentrated Work Scheduling: Enabling and Constraining Aspects." *Sociological Perspectives* 32: 47–64.

————. 1992. "Employment Data and Social Impact Assessment." *Evaluation and Program Planning* 15:1–7.

Gramling, Robert and Sarah Brabant, eds. 1984. *The Role of Outer Continental Shelf Activities in the Growth and Modification of Louisiana's Coastal Zone.* Lafayette, LA: U.S. Department of Commerce/Louisiana Department of Natural Resources.

———. 1986. "Boom Towns and Offshore Energy Impact Assessment: The Development of a Comprehensive Model." *Sociological Perspectives* 29:177–201.

Gramling, Robert and William R. Freudenburg. 1990. "A Closer Look at 'Local Control': Communities, Commodities, and the Collapse of the Coast." *Rural Sociology* 55 (4): 541–58.

———. 1992a. "Opportunity-Threat, Development, and Adaptation: Toward a Comprehensive Framework for Social Impact Assessment." *Rural Sociology* 57 (2, Summer): 216–34.

———. 1992b. "The Exxon Valdez Oil Spill In the Context of U.S. Petroleum Energy Politics." *Industrial Crisis Quarterly* 6 (3): 1–23.

Gramling, Robert and Edward Joubert. 1977. "The Impact of Outer Continental Shelf Petroleum Activity on Social and Cultural Characteristics of Morgan City, Louisiana." In E. F. Stallings, ed., *Outer Continental Shelf Impact, Morgan City, Louisiana,* 106–43. Baton Rouge: Louisiana Department of Transportation and Development.

Gulliford, Andrew. 1989. *Boomtown Blues: Colorado Oil Shale, 1885–1985.* Niwot, CO: University Press of Colorado.

Gundry, Kathleen G. and Thomas A. Heberlein. 1984. "Do Public Meetings Represent the Public?" *American Planning Association Journal* (Spring): 175–84.

Hance, Billie Jo, Caron Chess, and Peter M. Sandman. 1988. *Improving Dialogue with Communities: A Risk Communication Manual for Government.* New Brunswick, NJ: Rutgers University Environmental Communication Research Program.

Handy, C. 1978. *The Gods of Management.* London: Pan Books.

Heberlein, Thomas A. 1981. "Environmental Attitudes." *Zeitschrift für Umweltpolitik* 4 (February): 241–70.

Hershman, Arleen. 1977. "Regulating the Regulators." *Dun's Review* 109:34, 36.

Howell, Robert E., Darryll Olsen, Marvin E. Olsen, and Riley E. Dunlap. 1981. *Citizen Participation in Nuclear Waste Repository Siting.* Corvallis, OR: Western Rural Development Center.

Inkeles, Alex and David H. Smith. 1970. "The Fate of Personal Adjustment in the Process of Modernization." *International Journal Comparative Sociology* 11 (2, June): 81–114.

Jay, P. 1967. *Management and Machiavelli.* London: Hodder and Stoughton.

Jelienk, M. L., L. Smircich, and P. Hirsch, eds. 1985. "Organizational Culture." *Administrative Science Quarterly* 28: 123–36.

Johnson, James P. 1979. *The Politics of Soft Coal: The Bituminous Industry from World War I Through the New Deal.* Urbana, IL: Univ. of Illinois Press.

Jones, Lonnie L. 1988. "Adjustments to a Declining Resource Base: The Case of the Texas Economy." *Impact Assessment Bulletin* 6 (2): 81–91.

Jordan, William S. III. 1984. "Psychological Harm after PANE: NEPA's Requirements to Consider Psychological Damage." *Harvard Environmental Law Review* 8:55–87.

Kaplan, Elizabeth R. 1982. "California: Threatening the Golden Shore." In Joan Goldstein, ed., *The Politics of Offshore Oil*, 3–28. New York: Praeger.

Kaufman, Burton I. 1978. *Oil Cartel Case: A Documentary Study of Antitrust Activity in the Cold War.* Westport: Greenwood.

Killman, R. H., M. J. Saxton, and R. Serpa, eds. 1985. *Gaining Control of the Corporate Culture.* San Francisco: Jossey-Bass.

Krannich, Richard S., Thomas Greider, and Ronald L. Little. 1984. "Rapid Growth and Fear of Crime: A Four-Community Comparison." *Rural Sociology* 50 (2): 193–209.

Krauss, Celene. 1987. "Community Struggles and the State: From the Grassroots— A Practical Critique of Power." Presented at the American Sociological Association Annual Meeting, Chicago, August.

———. 1989. "Community Struggles and the Shaping of Democratic Consciousness." *Sociological Forum* 4 (2): 227–39.

Kunreuther, Howard, William H. Desvousges, and Paul Slovic. 1988. "Nevada's Predicament: Public Perceptions of Risk from the Proposed Nuclear Repository." *Environment* 30 (8, October): 16–20, 30–33.

Kunreuther, Howard, John Lathrop, and Joanne Linnerooth. 1982. "A Descriptive Model of Choice for Siting Facilities." *Behavioral Science* 27:281–97.

Lankford Raymond L. 1971. "Marine Drilling" In J. E. Brantley, ed., *History of Oil Well Drilling*, 1358–1444. Houston: Gulf Publishing Co.

Lantz, Alma E. and Robert L. McKeown. 1979. "Social/Psychological Problems of Women and Their Families Associated with Rapid Growth." In U.S. Commission on Civil Rights, eds., *Energy Resources Development: Implications for Women and Minorities in the Intermountain West*, 42–54. Washington, DC: U.S. Government Printing Office.

Laumann, Edward O. and David Knoke. 1987. *The Organizational State: Social Choice in National Policy Domains.* Madison, WI: University of Wisconsin Press.

Levine, Adeline G. 1982. *Love Canal: Science, Politics, and People.* Lexington, MA: Lexington.

Llewellyn, Lynn G. and William R. Freudenburg. 1990. "Legal Requirements for Social Impact Assessments: Assessing the Social Science Fallout from Three Mile Island." *Society and Natural Resources* 2 (3): 193–208.

Louisiana Department of Labor. 1970–87. Employment-Unemployment: Annual Average. Baton Rouge: Louisiana Department of Labor.

Lovejoy, Stephen B. and Richard S. Krannich. 1982. "Rural Industrial Development and Domestic Dependency Relations: Toward an Integrated Perspective." *Rural Sociology* 47 (Fall): 475–95.

Magnuson, W. G. and E. F. Hollings. 1975. *An Analysis of the Department of the Interior's Proposed Acceleration of Development of Oil and Gas on the Outer Continental Shelf.* Washington, D.C.: U.S. Government Printing Office.

Manuel, David P. 1980. "East St. Mary Parish in the 1970s: The Economics of a Sustained Energy Impact." In Robert Gramling, ed., *East St. Mary Parish, Economic Growth and Stabilization Strategies,* 44–48. Baton Rouge: Louisiana Department of Natural Resources.

———. 1984. "Trends in Louisiana OCS Activities." In Robert Gramling and Sarah Brabant, eds., *The Role of Outer Continental Shelf Activities in the Growth and Modification of Louisiana's Coastal Zone,* 27–40. Lafayette: U.S. Department of Commerce/Louisiana Department of Natural Resources.

———. 1985. "Unemployment and Drilling Activity In Major Energy-Producing States." *Journal of Energy and Development* 10:45–62.

Martin, Michael and Leonard Gelber. 1978. *The Dictionary of American History.* New York: Dorset Press.

McKay, Pat. 1988. "Politics of Oil Jams Fort Bragg: Coastal Towns Swell for Hearing." *Santa Rosa Press-Democrat* (Feb. 3):A1, A8.

Mead, W. J., A. Moseidjord, D. Mauraoka and P. Sorensen. 1985. *Offshore Lands: Oil and Gas Leasing and Conservation on the Outer Continental Shelf.* San Francisco: Pacific Institute for Public Policy Research.

Meidinger, Errol E. and William R. Freudenburg. 1983. "The Legal Status of Social Impact Assessments: Recent Developments." *Environmental Sociology* 34:30–33.

Milkman, Raymond H., Leon G. Hunt, William Pease, Una Perez, Lisa J. Crowley, and Brian Boyd. 1980. "Drug and Alcohol Abuse in Booming and Depressed Communities." Washington, DC: National Institute on Drug Abuse, U.S. Department of Health, Education and Welfare.

Miller, Ernest C. 1959. *Pennsylvania's Oil Industry.* Gettysburg, PA: Pennsylvania Historical Association (rev. ed.).

Mitchell, Robert C. 1979. "Silent Spring/Solid Majorities." *Public Opinion* 55(Aug/ Sept): 1–20.

———. 1984. "Rationality and Irrationality in the Public's Perception of Nuclear Power." In William R. Freudenburg and Eugene A. Rosa, eds., *Public Reactions to Nuclear Power: Are There Critical Masses?,* 137–79. Boulder, CO: American Association for the Advancement of Science/Westview.

Mohai, Paul. 1990. "Black Environmentalism." *Social Science Quarterly* 71 (4, Dec.): 744–65.

Molotch, Harvey. 1970. "Oil in Santa Barbara and Power in America." *Sociological Inquiry* 40 (Winter): 131–44.

———. 1976. "The City as a Growth Machine: Toward a Political Economy of Place." *American Journal of Sociology* 82:309–32.

Morgan City Historical Society. 1960. A History of Morgan City, Louisiana. Morgan City: Morgan City Historical Society.

Morgan, Gareth. 1986. *Images of Organizations.* London: Sage.

Morrison, Denton E. 1986. "How and Why Environmental Consciousness Has Trickled Down." In Allan Schnaiberg, Nicholas Watts and Klaus Zimmerman, eds., *Distributional Conflicts in Environmental-Resource Policy,* 187–220. New York: St. Martin's.

Mullins, Joe. 1981. "Sleepy Backwater Becomes . . . Boom Town, U.$.A." *National Enquirer* (Oct. 20): 9.

Murdock, Steve H. and F. Larry Leistritz. 1979. *Energy Development in the Western United States: Impact on Rural Areas.* New York: Praeger.

Nash, Gerald D. 1968. *United States Oil Policy 1890–1964.* Westport: Greenwood Press.

National Research Council. 1985. *Oil in the Sea: Inputs, Fates, and Effects.* Washington D.C.: National Academy Press.

National Research Council. 1989. *The Adequacy of Environmental Information for Outer Continental Shelf Oil and Gas Decisions: Florida and California.* Washington, DC: National Academy Press, National Academy of Sciences.

National Research Council. 1992. *Assessment of the U.S. Outer Continental Shelf Environmental Studies Program: III. Social and Economic Studies.* Washington, D.C.: National Academy Press, National Academy of Sciences.

National Research Council. 1993. *Assessment of the U.S. Outer Continental Shelf Environmental Studies Program: IV. Lessons and Opportunities.* Washington, D.C.: National Academy Press, National Academy of Sciences.

New York Times. 1975. "Pervious Industry Links Noted among U.S. Regulatory Aids." September 7: 36.

Nordhauser, Norman E. 1979. *The Quest for Stability: Domestic Oil Regulation 1917–1935.* New York: Garland.

Offe, Claus. 1984. *Contradictions of the Welfare State.* London: Hutchinson.

Oil and Gas Journal. 1988. Data Book. Tulsa: Pennwell Books.

Oil Weekly Staff. 1946. "Magnolia Company's Open Gulf Test Is Below 8400 Feet." *Oil Weekly,* October 28: 31–32.

Olson, Mancur. 1965. *The Logic of Collective Action.* Cambridge, MA: Harvard Univ. Press.

———. 1982. *The Rise and Decline of Nations: Economic Growth, Stagflation, and Social Rigidities.* New Haven: Yale University Press.

Pascale, T. J. 1982. *The Art of Japanese Management.* New York: Warner Books.

Paté-Cornell, M. Elisabeth. 1990. "Organizational Aspects of Engineering System Safety: The Case of Offshore Platforms." *Science* 250 (Nov. 30): 1210–17.

Paté-Cornell, M. Elisabeth and Robert G. Bea. 1992. "Management Errors and System Reliability: A Probabilistic Approach and Application to Offshore Platforms." *Risk Analysis* 12 (1, March): 1–18.

Perrow, Charles. 1984. *Normal Accidents: Living with High-Risk Technologies.* New York: Basic.

Peters, T. J. 1978. "Symbols, Patterns and Settings." *Organizational Dynamics* 7:3–22.

Popper, Frank J. 1981. "Siting LULUs." *Journal of the American Planning Association* 47 (4, April): 12–15.

Pressman, Jeffrey L. and Aaron Wildavsky. 1973. *Implementation: How Great Expectations in Washington are Dashed in Oakland; or, Why It's Amazing that Federal Programs Work at All, This Being a Saga of the Economic Development Administration as told by two Sympathetic Observers who seek to build Morals on a Foundation of Ruined Hopes.* Berkeley: University of California Press.

Pring, G. W. 1981. " 'Power to Spare': Conditioning Federal Resource Leases to Protect Social, Economic, and Environmental Values." *Natural Resources Lawyer* 14 (2): 305–38.

Rayner, Steve and Robin Cantor. 1987. "How Fair is Safe Enough? The Cultural Approach to Technology Choice." *Risk Analysis* 7 (1, March): 3–9.

Real Estate Research Corporation. 1974. *The Costs of Sprawl: Detailed Cost Analysis.* Washington, DC: U.S. Council on Environmental Quality.

Reilly, T. F. 1980. "Recreation." In Robert Gramling, ed., *East St. Mary Parish, Economic Growth and Stabilization Strategies*, 288–302. Baton Rouge: Louisiana Department of Natural Resources.

Royal Commission on the Ocean Ranger Marine Disaster. 1984. *The Loss of the Semisubmersible Drill Rig Ocean Ranger and its Crew.* Ottawa: Canadian Government Printing Centre.

Sampson, Anthony. 1975. *The Seven Sisters: The Great Oil Companies and the World They Made.* New York: Viking.

Savatsky, Pamela D. 1974. "A Legal Rationale for the Sociologist's Role in Researching Social Impacts." In C. P. Wolf, ed., *Social Impact Assessment*, 45–47. Stroudsburg, PA: Dowden, Hutchinson & Ross.

Schein, E. 1982. *Organizational Culture and Leadership.* San Francisco: Jossey-Bass.

Schnaiberg, Allan. 1980. *The Environment: From Surplus to Scarcity.* New York: Oxford University Press.

Selznick, Philip. 1948. "Foundations of the Theory of Organization." *American Sociological Review* 13:25–35.

Seyfrit, Carole L. 1986. "Migration Intentions of Rural Youth: Testing an Assumed Benefit of Rapid Growth." *Rural Sociology* 51 (2, Summer): 199–211.

Slovic, Paul. 1987. "Perception of Risk." *Science* 236:280–85.

Slovic, Paul, Baruch Fischhoff and Sarah Lichtenstein. 1984. "Perception and Acceptability of Risk from Energy Systems." In William R. Freudenburg and Eugene A. Rosa, eds., *Public Reactions to Nuclear Power: Are There Critical Masses?*, 115–35. Boulder, CO: American Association for the Advancement of Science/Westview.

Slovic, Paul, James H. Flynn, and Mark Layman. 1991. "Perceived Risk, Trust, and the Politics of Nuclear Waste." *Science* 254 (13 December): 1603–07.

Smircich, E. 1983. "Concepts of Culture and Organizational Analysis." *Administrative Science Quarterly* 28:339–58.

Stallings, E. F. 1984. "Development of OCS Techniques." In Robert Gramling and Sarah Brabant, eds., *The Role of Outer Continental Shelf Activities in the Growth and Modification of Louisiana's Coastal Zone*, 1–26. Lafayette: U.S. Department of Commerce/Louisiana Department of Natural Resources.

Stallings, E. F. and T. F. Reilly. 1980. "Transportation, East St. Mary Parish." In Robert Gramling, ed., *East St. Mary Parish, Economic Growth and Stabilization Strategies*, 278–88. Baton Rouge: Louisiana Department of Natural Resources.

Stallings, Robert. 1990. "Media Discourse and the Social Construction of Risk." *Social Problems* 37 (1, Feb.): 80–95

Stanley, David. Forthcoming. "Hydro-Acoustic Assessment of Abundance and Behavior of Fish Associated with Oil and Gas Platforms of the Louisiana Coast." *Bulletin of Marine Science.*

Stone, Clarence N. 1980. "Systemic Power in Community Decision Making: A Restatement of Stratification Theory." *American Political Science Review* 74:978–90.

Summers, Gene F., Sharon D. Evans, Frank Clemente, Elwood M. Beck, and Jon Minkoff. 1976. *Industrial Invasion of Nonmetropolitan America: A Quarter Century of Experience.* New York: Praeger.

Szasz, Andrew. 1990. "From Pollution Control to Pollution Prevention: How Does It Happen?" Presented to annual meeting of American Sociological Association, Washington, DC, August.

Tarbell, Ida M. 1904. *The History of the Standard Oil Company.* New York: Macmillan.

Turner, R. Eugene and Donald R. Cahoon. 1988. *Causes of Wetlands Loss in the Coastal Central Gulf of Mexico.* New Orleans: U.S. Minerals Management Service.

United States V. California, 332 U.S. 19 (1947).

U.S. Council on Environment Quality. 1978. *Regulation for Implementing the Procedural Provisions of the National Environmental Policy Act* (40 CFR 1500–08). Washington, DC: U.S. Council on Enviornmental Quality.

U.S. Department of Commerce, Bureau of the Census. 1940. *1940 Census of Population: Characteristics of the Population.* Washington, DC: U.S. Government Printing Office.

———. 1970. *1970 Census of Population: Characteristics of the Population.* Washington, DC: U.S. Government Printing Office.

———. 1980. *1980 Census of Population: Characteristics of the Population.* Washington, DC: U.S. Government Printing Office.

U.S. Department of Interior, U.S. Department of Agriculture, and U.S. Interstate Commerce Commission. 1974. *Final Environmental Impact Statement of Proposed Coal Development in the Eastern Powder River Basin of Wyoming.* Washington, DC: U.S. Department of Interior.

U.S. Energy Information Administration. 1992. "U.S. Oil and Gas Reserves Declining." *Energy Information Administration New Releases* (Sept.–Oct.): 1, 2.

U.S. Federal Trade Commission. 1952. *The International Petroleum Cartel.* Washington, D.C.: U.S. Government Printing Office.

U.S. Minerals Management Service. 1987. *Proposed 5-Year Outer Continental Shelf Oil and Gas Leasing Program, Mid-1987 to Mid-1992: final Environmental Impact Statement.* Reston, VA: Minerals Management Service.

U.S. Minerals Management Service. 1988. *Oil and Gas Program: Cumulative Effects.* Washington, DC: U.S. Department of the Interior (MMS 88-0005).

————. 1989. *OCS National Compendium.* Washington, DC: U.S. Minerals Management Service, U.S. Department of the Interior.

————. 1991. *Comprehensive Program 1992–1997: Draft Environmental Impact Statement.* "Executive Summary," pp. i–viii.

Van Liere, Kent D. and Riley E. Dunlap. 1980. "The Social Bases of Environmental Concern: A Review of Hypotheses, Explanations and Empirical Evidence." *Public Opinion Quarterly* 44:181–97.

————. 1981. "Environmental Concern—Does it Make a Difference How It's Measured?" *Environment and Behavior* 13 (November): 651–76.

Waataja, Ron, 1988. "Activists Hope For Invasion By Land," *North Coast News* 2 (#8, Jan. 21–Feb. 3): 1.

Webber, R. A., ed. 1969. *Culture and Management.* Homewood, IL: Irwin.

Weber, Bruce A. and Robert E. Howell. 1982. *Coping with Rapid Growth in Rural Communities.* Boulder, CO: Westview.

West, Patrick C. 1982. *Natural Resource Bureaucracy and Rural Poverty: A Study in Political Sociology of Natural Resources.* Ann Arbor, MI: University of Michigan School of Natural Resources.

Wilkins, Mira 1976. "The Oil Companies in Perspective." In Raymond Vernon, ed., *The Oil Crisis,* 159–78. New York: Norton.

Wilkinson, Kenneth P., James G. Thompson, Robert R. Reynolds, Jr. and Laurence M. Ostresh. 1982. "Local Social Disruption and Western Energy Development: A Critical Review." *Pacific Sociological Review* 25:275–96.

Wilson, E. 1982. "MAGCRC: A Classic Model for State/Federal Communication and Cooperation." In Joan Goldstein, ed., *The Politics of Offshore Oil,* 72–86. New York: Praeger.

World Resources Institute. 1993. *The 1993 Information Please Environmental Almanac.* Boston: Houghton Mifflin.

Wybrow, Peter. 1986. "Comparative Responses and Experiences to Migration Due to Oil Development in Scotland." In ISER Conference Papers No. 1. St. John's, Newfoundland: Institute of Social and Economic Research, Memorial University of Newfoundland.

Wynne, Brian. 1982. *Rationality and Ritual: The Windscale Inquiry and Nuclear Decisions in Britain.* Chalfont St. Giles: British Society for the History of Science.

NOTES

CHAPTER 2

1. The name comes from the Achnacarry Castle in Scotland, where the leaders of the major corporations met to hammer out the agreement. Also known as the "As-is" agreement, the pact effectively divided up the world's oil markets among major oil companies and provided that the prices for oil, anywhere in the world, would be based on what the oil would have cost if it had been shipped there from the Gulf of Mexico (for further discussion, see U.S. Federal Trade Commission 1952; Engler 1961).

2. For a brief but more thorough examination of this and other relevant federal legislation, plus a brief discussion of how to interpret the legal citations, see the appendix. Even in combination, of course, these two laws did not settle all of the existing issues. In particular, conflicts continued for years over the leases that were in federal waters but sufficiently close to the state waters to raise the argument that the federal leases were also draining oil under state-controlled waters. When the issue was finally settled in 1978, the states were given 27% of the royalties from leases that extended an additional three miles offshore.

3. Federal revenue from OCS leases comes from two sources. A bonus bid is a sealed, theoretically competitive bid offered by a company, or group of companies, to secure the acreage. Royalties represent a percentage of the profit from the exploitation of any oil that is actually discovered and extracted.

4. "P.L." stands for "Public Law." For a more detailed discussion of legal citation formats, see the appendix.

5. For readers who are interested, the appendix also provides a discussion of how to interpret legal citations such as this one.

CHAPTER 3

6. This is a reference to the "concentrated" work scheduling discussed in the previous chapter, where a worker puts in a full week of 12-hour days and then is given seven days off.

7. Many of the coastal regions have been criss-crossed by access canals that have been dug to and from drilling locations, as well as by an extensive maze of pipelines to serve the onshore as well as offshore rigs.

8. PIRO stands for the Petroleum Industry Response Organization, which was set up by petroleum companies after the *Exxon Valdez* accident to help deal with oil spills.

The organization has since been replaced by the Marine Spill Response Corporation, or MSRC.

9. Several residents noted that in Eureka's most recent city elections at the time of the interviews, where the mayor's office and three of the five council seats were being contested, and where all of these positions had previously been occupied by supporters of offshore oil development, all of the elections were won by candidates who were opposed to OCS development. Of the two remaining pro-OCS city council members, one was facing a potential recall at the time of the interviews.

10. A representative from a local governmental body, for example, might express concern about the congestion or the wear and tear that OCS-related traffic could create for her city's streets, while a member of the general public from the same community might be more likely to express concerns that "offshore oil development," in general, could create impacts that the citizen would find objectionable.

CHAPTER 4

11. The rationale for selecting these four goals is not immediately apparent. Our best guess is that, for reasons that are not entirely clear, the leaders of MMS chose to emphasize not the act's own listing of policy goals in section 1332, but four subpoints listed under the second of the ten main points of an entirely different section of the Act (43 U.S.C. 1802(2)). That second main point calls on the Secretary of Interior "to preserve, protect, and develop oil and natural gas resources . . . in a manner which is consistent with" essentially the four sub-points listed. Some of the other *main* points of that same section of the law, however, call for the development of new and improved technology to "eliminate or minimize risk of damage to the human, marine, and coastal environments," to provide states with information and other "comprehensive assistance . . . [to] assure adequate protection of the human environment," and to "minimize or eliminate conflict" between oil and gas activities "and the recovery of other resources such as fish and shellfish" (43 U.S.C. 1802 (3), (4), (5), (6), and (7)). What makes this choice somewhat more confusing is that the 1800 "chapter" is *not* the location of the act's discussion of "major goals," as MMS's Mission Statement claims; major management goals are listed in chapter 1300. Instead, chapter 1800 contains three subchapters—one establishing an "Offshore Oil Spill Pollution Fund," a second establishing a "Fisherman's Contingency Fund," and a third setting forth "Miscellaneous Provisions" (43 U.S.C. 1801 et seq.). For further discussion of legal requirements, with particular emphasis on the implications for studies of the human environment, see the appendix.

APPENDIX

1. For those who are utterly confused by legal citations, this note will provide a simplified, non-lawyers' summary. The numbering starts when Congress passes a law; at this point, a law is commonly given a *P.L.* ("Public Law") number. P.L. 94–140, for example, is essentially law number 140 passed by the 94th Congress. This numbering system is reasonably straightforward, but it has nothing to do with the con-

tent of the law, so a second system has been developed—the *U.S.C.* ("U.S. Code") system, which recategorizes the laws in terms of broad content categories—such that 42 U.S.C. 4321 et seq., for example, refers to volume 42 of the U.S. Code, part 4321 "and following." (Subsequent actions by Congress can increase or decrease the number of following parts.)

Given that a single piece of legislation can often deal with more than just one topic, it is entirely possible that different pieces of the law will wind up being classified at very different spots in the U.S. Code. The usual tendency is to refer to a law in terms of its "P.L." number at the time when it is first passed, which is also the time when the law usually gets the most attention from the public as a whole, but in general, by the time most lawsuits make their way to court, the "U.S.C." numbers have already been assigned.

Finally, as illustrated by the discussion of bureaucratic slippage in chapter 5, the laws passed by Congress tend to be quite broad and vague; the more specific "regulations" to "implement" these laws or statutes are normally "promulgated" (to use the common jargon) by federal agencies. These regulations have a third numbering system, one that is entirely their own—*C.F.R.* ("Code of Federal Regulations") numbers. The regulations implementing the procedural provisions of the National Environmental Policy Act, for example, are found at 40 C.F.R. 1500 et seq.—volume 40 of the Code of Federal Regulations, parts 1500 and following.

INDEX